THE AMPLIFIED MIND

THE AMPLIFIED MIND

THINKING CLEARLY IN THE AGE OF AI

PAUL F. ALDEN

ABOOKS
Alive Book Publishing

The Amplified Mind
Thinking Clearly in the Age of AI

Additional copies may be ordered from the publisher for educational, business, promotional or premium use.
For information, contact ALIVE Book Publishing at:
alivebookpublishing.com

Book Design by Alex Johnson

ISBN 13
978-1-63132-285-3

Library of Congress Control Number: 2026915453

Library of Congress Cataloging-in-Publication Data
is available upon request.

First Edition

Published in the United States of America by
ALIVE Book Publishing
an imprint of Advanced Publishing LLC
3200 A Danville Blvd., Suite 204, Alamo, California 94507
alivebookpublishing.com

PRINTED IN THE UNITED STATES OF AMERICA

10 9 8 7 6 5 4 3 2 1

For my family, Leslie, AJ, and Delaney,
and for Oakley, who keeps me company
through every page. And in loving memory
of my mother and father, who gave me
my first lessons in critical thinking.

CONTENTS

Preface 13

The Sequence That Changed

Introduction 17

Agency in an Age of Synthetic Fluency

PART I: THE DIAGNOSTIC ARC

Chapter 1 Fluency Distortion 25

When Clarity Replaces Understanding

Chapter 2 Cognitive Offloading 33

When Thinking Quietly Moves Outside the Mind

Chapter 3 Authority Transfer 43

When Approval Replaces Origination

Chapter 4 Reactive Minds 51

When Generation Becomes Reaction

Chapter 5 The Strain 61

The Cost of Convenience

Chapter 6 Authorship 71

When Editing Replaces Authoring

Chapter 7 The Signal Problem 81

Protecting Attention from Useful Interruption

Chapter 8 Institutional Convergence 89

When Individual Deference Becomes Collective Uniformity

Chapter 9 Knowing Versus Accessing 99

What the Mind Must Still Hold

PART II: THE CONSTRUCTIVE ARC

Chapter 10 Analysis Versus Judgment 109

Where Human Reasoning Remains Essential

Chapter 11 The Amplified Mind 119

Substitution Versus Amplification

Chapter 12 Designing Amplified Workflows 129

Accidental Versus Intentional Design

Chapter 13 Designing Organizational Intelligence 141

Synchronized Thinking Versus Collective Intelligence

Chapter 14 The Discipline of Amplified Thinking 153

The Choice That Appears Invisible

Notes **161**

Glossary **169**

Index **177**

FOREWORD

By Professor Grant Schofield PhD
Professor of Public Health,
Auckland University of Technology;
Director, Human Potential Centre;
Chief Science Officer, PREKURE

I came to Paul Alden's argument from the wrong door, which is the best way to arrive at an idea. My field is public health and human metabolism. His is cognition. We have spent years on opposite sides of the same building without knowing the other was inside.

The thesis we both ended up holding is this. Every technological revolution that solves a problem for the human body or mind also removes a stimulus that body or mind was built to need. Walking became driving. Hunger became continuous ultraprocessed food. Hard physical work became sitting at a screen. Boredom was replaced by continuous entertainment through scrolling. Each of those revolutions improved life on its own terms. Each one also produced a new shape of illness that did not exist before, because the friction had been carrying information the organism needed to stay calibrated. This is what evolutionary biologists call mismatch. Daniel Lieberman has documented the physical version of it carefully in *The Story of the Human Body*. Paul has now documented the cognitive version, and he has done it with the same seriousness.

Cognitive offloading is not new. Plato worried about writing. We have had robust evidence since Sparrow and colleagues published in Science in 2011 that people remember less when they expect a search engine to remember for them. London taxi drivers grew measurably larger posterior hippocampi from learning the streets the hard way, and there are now signals in the literature that the opposite happens when navigation is outsourced to a phone. What is new with large language models is the scope. We are not offloading one cognitive function. We are offloading the act of thinking itself, including the slow, frustrating, productive struggle that builds judgment.

This is the friction Paul writes about. And here is where I want readers to pay close attention. The book does not argue that AI is bad. It argues that AI is frictionless, and that frictionless is a cognitive condition the human nervous system has never had to handle before. Consider what that actually means. For the entirety of human evolutionary history, thinking has required effort. Forming an argument, constructing a narrative, working through a problem without a ready answer, these were the activities that built and maintained the neural architecture of judgment. We are now the first generation to have that effort made optional on a mass scale. We do not yet know what happens to a mind that never has to struggle. But the metabolic evidence suggests we should be paying very close attention, because organisms that lose the stimuli that shaped them do not simply stay the same. They change, quietly and cumulatively, in the direction the new environment selects for.

The same way we now have to deliberately reintroduce movement, hunger, cold, and discomfort to stay metabolically healthy, we will have to deliberately reintroduce cognitive effort to stay mentally sharp. The question is no longer whether to use these tools. The question is how to use them without quietly letting the parts of being human that matter most fade away.

What I find genuinely useful about *The Amplified Mind* is that it does not stop at the diagnosis. Most books in this space identify the problem and leave the reader to figure out what to do on a Tuesday morning. Paul has built something practical. The five minute rule alone is worth the price of the book. It is the kind of simple, repeatable, self-administered discipline that real behaviour change actually needs. It is the cognitive equivalent of a daily walk or a skipped meal. Small, structured friction, deliberately reintroduced.

I want to say something directly about who needs this book most. Not the fifty year old professional who built their thinking before these tools existed and is now learning to use them carefully. That person has the cognitive architecture already in place. I am thinking about the student sitting in a classroom right now, or about to enter one, for whom AI has always been available, for whom the struggle of an unaided first draft is already beginning to feel unnecessary. Their creativity, their independent voice, their capacity for original thought, these are still forming. The mismatch is happening in real time for them in a way it simply is not for those of us who earned our thinking the harder way. The sooner we place this argument in their hands, the better chance they have of understanding what is quietly at stake and choosing accordingly.

Read this book carefully. Then run his suggestions on yourself. You will be a better thinker for the inconvenience.

Grant Schofield PhD

Professor of Public Health,
Auckland University of Technology;
Director, Human Potential Centre;
Chief Science Officer, PREKURE

PREFACE

The issue is not whether artificial intelligence improves thinking or degrades it. The issue is how the structure of thinking changes when intelligent systems become part of the reasoning process itself.

For decades the public conversation about technology and cognition has followed a familiar pattern. Each major shift in information tools raises the same concern: that our intellectual capacities may be weakening as our tools grow more capable.

Nicholas Carr argued that the internet altered the conditions of attention and concentration. The problem he described was not intelligence but distraction, the fragmentation of focus in an environment designed to deliver constant streams of information. Cal Newport extended that concern into the professional world, showing how knowledge work suffers when deep thinking is replaced by rapid digital responsiveness. Daniel Kahneman approached the problem from a different direction, demonstrating how human judgment is systematically distorted by cognitive shortcuts and biases.

Each of these analyses describes a genuine transformation in the way people think. But artificial intelligence introduces a different change, one that does not fit neatly into any of these frameworks.

The internet flooded the mind with information. Artificial intelligence generates reasoning.

This shift alters the sequence through which thinking happens.

For most of modern history the order of cognition was stable. A person generated an idea, a hypothesis, or a line of reasoning internally. Tools then assisted in evaluating, testing, or refining that reasoning. Calculators checked arithmetic. Databases retrieved facts. Software accelerated analysis. The machine supported the thinking process, but the thinking began with the human mind.

Artificial intelligence reverses that sequence. The system generates a structured answer first. The human evaluates afterward.

That reversal appears efficient, and in many situations it is. But it quietly changes the cognitive role the human occupies in the reasoning process. Instead of constructing arguments or explanations from first principles, the human becomes the evaluator of a generated result. The difference seems subtle in the moment, yet over time it alters how expertise develops, how judgment forms, and how professionals understand the problems they face.

I began noticing this shift not in research papers but in practice.

For most of my professional life I worked in environments where decisions carried immediate and tangible consequences. As a firefighter and later in operational and technical roles connected to infrastructure and complex systems, I spent decades watching how professionals think when information is incomplete, conditions are uncertain, and mistakes have real costs. In those environments, judgment is not an abstract intellectual exercise. It is the difference between control and chaos.

In recent years I began seeing the same decision environments intersect with rapidly advancing intelligent systems. Organizations were adopting artificial intelligence to assist with planning, analysis, and operational decisions. The technology was powerful and often extraordinarily useful. But something subtle was happening to the reasoning habits of the people using it. The tools were not only accelerating work. They were beginning to shape the sequence through which thinking itself occurred.

People who once constructed explanations were increasingly evaluating generated ones.

The consequences of that shift were rarely visible immediately. In many cases the systems produced correct answers. The danger appeared only later, when unfamiliar problems arose and the human operator no longer possessed the internal understanding needed to recognize whether the system was right.

This book is an attempt to describe that emerging change clearly.

The chapters that follow examine how artificial intelligence produces what can be called *synthetic fluency*, responses that resemble well-formed reasoning even when the underlying logic may be incomplete or incorrect. They explore how reliance on these systems alters professional judgment, redistributes intellectual authority, and gradually reshapes the boundary between assistance and dependency.

Understanding that transformation is the first task of this book.

The second is practical. If intelligent systems are becoming permanent partners in human reasoning, the challenge is not to avoid them but to design ways of working with them that preserve the human capacity to think independently. The goal is not to reject artificial intelligence, nor to embrace it uncritically, but to understand how the partnership between human judgment and machine generation can be structured so that each strengthens the other.

The difference between those two approaches will not appear immediately.

It will appear later, when complex problems arise that cannot be solved by generated answers alone, and when the people facing those problems must decide whether they still possess the understanding required to think beyond the system in front of them.

INTRODUCTION

AGENCY IN AN AGE OF SYNTHETIC FLUENCY

A professional sits down to begin thinking.

The task might be a strategy memo, a research brief, a legal argument, or a market analysis. For generations the process began the same way. A blank page appeared, and the first structure of thought had to come from the mind of the person responsible for the work.

The opening idea might be incomplete.
The structure might be wrong.
The reasoning might require revision.

But the architecture of the problem belonged to the thinker.

Artificial intelligence changes that starting point.

Today the same professional opens a system and types a prompt. Within seconds a structured argument appears. The system produces a clean outline, supporting points, and a coherent explanation of the issue at hand. The language is clear. The reasoning sounds persuasive. The structure appears thoughtful.

The work has begun.

But something fundamental has changed.

For the first time in the history of intellectual tools, the **initial structure of thought no longer has to originate in the mind of the person responsible for the decision**.

The system generates it.

What makes this shift powerful is not simply speed. It is the quality of the output. Artificial intelligence produces language that feels immediately coherent. Arguments arrive already organized. Explanations appear fluent before the human mind has fully engaged with the problem.

This quality can be described as **synthetic fluency**.

Fluency has always been persuasive. When information is easy to process, it feels more credible. When reasoning arrives clearly structured, it feels more trustworthy. Human judgment is naturally influenced by the effort required to understand an idea. The easier something feels to process, the more convincing it often appears.

Artificial intelligence now produces that fluency instantly.

The result is that the **first structure of thought increasingly comes from the system rather than the thinker**.

At first glance the difference seems trivial.

Evaluation still occurs. Decisions are still made. Reports are still written. The human remains responsible for the outcome.

Nothing appears broken.

Yet the sequence of thinking has quietly changed.

For centuries intellectual work followed a stable order.

Humans generated ideas.
Humans evaluated those ideas.
Tools assisted along the way.

Artificial intelligence often reverses that order.

The system generates first.

The human evaluates afterward.

That reversal matters because expertise does not develop through evaluation alone.

Experts are not formed by reviewing ideas.

Experts are formed by **struggling to produce them**.

The effort required to construct an explanation, organize an argument, or frame a complex problem forces the mind to build internal structures. These structures allow experienced thinkers to recognize patterns, diagnose ambiguity, and respond intelligently when conditions are unclear.

They form slowly.

They form through repeated acts of generation, correction, and refinement.

When the generative step moves outside the mind, the conditions that build those structures begin to change.

This does not mean artificial intelligence inevitably weakens thinking.

The reality is more complex.

Artificial intelligence introduces a divergence in how professionals work.

Some individuals will begin thinking with the system. The machine produces the initial structure. The human reviews, edits, and refines. Work becomes faster and more fluent, but the internal architecture that produces ideas may gradually weaken.

Others will adopt a different sequence. They will generate their own models first and use artificial intelligence to test assumptions, challenge reasoning, and accelerate refinement.

Both groups will appear productive.

Both will produce well-structured outputs.

But over time their cognitive architecture will diverge.

One group becomes increasingly dependent on systems to supply the structure of reasoning.

The other learns to use those systems to **amplify thinking that still originates in the human mind**.

The difference will not be obvious in everyday productivity.

It will become visible under pressure.

When ambiguity is high.
When information is incomplete.
When time compresses and decisions must be defended publicly.

In those moments, judgment depends on internal architecture.

The ability to frame problems, weigh competing explanations, and tolerate uncertainty cannot be outsourced entirely. It depends on cognitive structures that exist before the machine produces an answer.

Artificial intelligence does not eliminate those abilities.

But it changes the conditions under which they develop.

At its core this book is about **agency**.

In a professional context, agency is the capacity to construct an initial model of a problem before external systems supply one, to evaluate arguments independently rather than accepting fluent output at face value, and to take responsibility for decisions because the reasoning behind them is internally understood.

Artificial intelligence does not automatically remove that capacity.

What it changes is the starting point of thinking.

When systems generate structure first and humans evaluate second, the cognitive burden shifts. Over time that shift can alter how problems are framed, how effort is allocated, and where responsibility ultimately resides.

Nothing appears broken in the moment.

Work still gets done.
Deadlines are still met.
Performance metrics may even improve.

But the architecture that produced those outputs may no longer belong primarily to the human operator.

This dynamic is part of a broader pattern in human cognition. Throughout history, people have extended their mental abilities through external tools. Writing expanded memory. Calculators expanded arithmetic. Digital search expanded information retrieval.

Artificial intelligence expands the process further.

It does not simply store or retrieve knowledge.

It generates explanations, arguments, summaries, and strategies. It produces reasoning that appears organized and authoritative the moment it appears.

In an environment where structured language becomes universally accessible, surface competence begins to equalize.

Clean reports, executive summaries, and strategic outlines are no longer scarce outputs. Anyone with access to the same systems can produce them.

When fluency becomes universal, differentiation moves elsewhere.

It moves to depth.

It moves to judgment under uncertainty.

It moves to sustained attention when complexity resists simplification.

These are not features of software.

They are features of disciplined cognition.

The professionals who retain the ability to generate structure before consulting systems will operate differently from those who default to system-generated framing.

The difference will not be obvious during routine work.

It will emerge when decisions carry real consequences.

Artificial intelligence is often described as a tool that will make knowledge workers more efficient.

That description is incomplete.

Artificial intelligence is reorganizing the architecture of thinking itself.

The central question is not whether machines will produce reasoning.

They already do.

The question is whether humans will continue to develop the internal structures required to think well in a world where convincing reasoning can be generated instantly.

The chapters that follow examine how this shift unfolds and how professionals can adapt without surrendering authorship of thought.

The goal is not to reject artificial intelligence.

The goal is to use it without losing the cognitive architecture that makes judgment possible.

In the emerging cognitive landscape, the most valuable professional skill will not be prompt writing.

It will be **maintaining authorship of thought in an age of synthetic fluency**.

CHAPTER 1

FLUENCY DISTORTION

WHEN CLARITY REPLACES UNDERSTANDING

A strategy meeting begins at nine.

The executive team has gathered to review a proposed acquisition. The target company is smaller but promising. The technology appears valuable. The financial projections are attractive, but the integration risks are unclear.

Someone has prepared a briefing document.

The analysis is thorough. The structure is clean. Each section follows logically from the last. Risks are acknowledged and categorized. The language is precise, the reasoning appears balanced, and the conclusions feel measured.

The document is persuasive.

Questions are asked. A few numbers are clarified. The group moves forward.

The meeting ends.

What no one in the room noticed was how little scrutiny the document actually received.

Not because the people involved lacked expertise.

Not because the analysis was necessarily wrong.

But because the document felt easy to understand.

Its clarity created the impression that the reasoning behind it had already been tested.

The explanation felt complete.

The structure felt convincing.

And that feeling subtly reduced the impulse to examine it more deeply.

This phenomenon can be described as **fluency distortion**.

Fluency distortion occurs when the **ease with which an explanation can be processed is mistaken for evidence that the explanation is correct**.

The clearer an argument appears, the more convincing it often feels.

But clarity and correctness are not the same variable.

An explanation can be structured elegantly while the reasoning beneath it remains incomplete.

A conclusion can sound confident while the evidence supporting it is thin.

Language can appear polished even when the underlying thinking has not been fully tested.

Psychologists studying judgment have documented this tendency for decades. Research on what is known as **processing fluency** shows that information which feels easier to process is more likely to be judged as true, regardless of its actual accuracy.

The mind interprets ease as reliability.

Artificial intelligence amplifies this effect.

For the first time, highly fluent explanations can be generated instantly, regardless of whether the user producing them fully understands the reasoning involved.

The result is a subtle but powerful shift in how ideas are evaluated.

To understand why this happens, it helps to recognize a simple principle of cognition:

Friction stimulates scrutiny.

When an explanation is difficult to follow, people naturally slow down.

They reread sentences.

They question assumptions.

They test whether the argument actually holds together.

Cognitive friction activates evaluation.

Fluency has the opposite effect.

When an explanation feels easy to process, the mind tends to accept it more quickly. The smoothness of the language creates the impression that the reasoning has already been organized and verified.

Instead of examining the structure behind the argument, the mind moves forward.

For most of human history this bias operated within natural limits. Producing fluent explanations required effort. Clear language usually implied that someone had invested the time necessary to think carefully about the issue.

Artificial intelligence breaks that relationship.

Clarity can now appear before deep reasoning has occurred.

Structure can be generated without genuine understanding.

Confidence can emerge even when the underlying logic remains incomplete.

This does not mean that AI-generated explanations are unreliable.

Many are extremely useful.

The risk is more subtle.

Fluency can reduce the level of scrutiny applied to the reasoning behind an explanation.

Artificial intelligence introduces this distortion through three interacting forces.

The first is **processing fluency**. Information that is easy to process feels more credible.

The second is **fast intuitive acceptance**. When explanations appear coherent, the mind often accepts them before engaging slower analytical reasoning.

The third is **machine-generated fluency**. Artificial intelligence dramatically increases the availability of highly structured explanations, making fluent reasoning ubiquitous.

Together these forces produce a new cognitive environment.

Professionals are now surrounded by explanations that appear immediately convincing.

The challenge is not that these explanations are always wrong.

The challenge is that their **clarity can exceed the depth of scrutiny applied to them**.

This distortion becomes most dangerous in decision environments where mistaken reasoning carries real consequences.

Strategic acquisitions.

Legal risk assessments.

Medical diagnoses.

Financial forecasting.

Policy design.

These decisions depend not only on receiving information but on interrogating the assumptions that produced it.

Fluency compresses that interrogation.

When an explanation appears complete, the mind assumes the difficult thinking has already occurred.

But appearance is not proof.

Clarity is not verification.

Persuasion is not evidence.

The most effective defense against fluency distortion is not rejecting fluent explanations.

It is restoring friction deliberately.

Friction slows thinking down.

It forces the mind to examine structure rather than accepting it automatically.

Professionals can introduce this friction through simple practices.

One is the **clarity test**.

When encountering a polished explanation, ask a simple question:

Would this argument still feel convincing if it were written poorly?

Imagine the same reasoning expressed awkwardly, without structure or rhetorical polish. Would the logic still hold? Would the evidence still appear sufficient?

If the argument loses its persuasive power when its language becomes less elegant, fluency may be doing more work than reasoning.

A second practice is **structural reconstruction**.

Instead of accepting the explanation as presented, attempt to rebuild the argument independently. What assumptions does it rely on? What evidence connects the steps? Where could alternative interpretations exist?

Reconstruction restores the generative effort that fluent explanations often remove.

A third practice is **friction insertion**.

Before accepting a persuasive analysis, introduce a deliberate pause. Ask what evidence would contradict the conclusion. Ask what uncertainties the explanation may have compressed. Ask whether the reasoning would survive outside the clean structure in which it was presented.

These practices do not reject artificial intelligence.

They simply counteract the cognitive shortcut that fluency triggers.

The goal is not to distrust clear explanations.

The goal is to ensure that clarity does not replace understanding.

Because fluency distortion rarely appears as an obvious mistake.

It appears as a feeling.

The explanation makes sense.

The reasoning feels complete.

The conclusion appears obvious.

And that feeling is often enough to move forward.

The real danger of fluency distortion is not that it produces bad explanations.

It is that **good explanations stop being examined once they sound convincing enough**.

Exercises

Exercise 1: The Clarity Test

Take a polished explanation produced by an AI system.

Rewrite it deliberately in rough, unstructured language. Remove bullet points. Break the logical flow.

Now ask yourself: does the argument remain convincing, or did its persuasive power depend on its structure?

Exercise 2: Friction Reconstruction

Read an AI-generated explanation of a complex issue.

Close the document.

Attempt to reconstruct the reasoning independently from memory. What assumptions did the argument rely on? What evidence supported the conclusion?

Compare your reconstruction to the original explanation. Notice where the reasoning felt solid and where clarity may have masked uncertainty.

Exercise 3: Friction Insertion

Before accepting a fluent explanation, pause and ask three questions:

What assumption does this argument rely on most heavily?

What evidence would contradict the conclusion?

What alternative explanation could also fit the facts?

These questions introduce the friction that fluent language often removes.

CHAPTER 2

COGNITIVE OFFLOADING

WHEN THINKING QUIETLY MOVES OUTSIDE THE MIND

The analyst began with a familiar kind of problem.

A client had asked whether a proposed regulatory change might expose the company's supply chain to new compliance risks. It was the type of question analysts encounter constantly: a new rule appears, the language is dense, and the operational implications are not immediately obvious.

Ordinarily the work would begin with the regulation itself. Identify the relevant provisions. Trace how they interact with existing transportation rules. Estimate the operational consequences.

Instead, the analyst opened an artificial intelligence interface.

"Explain the likely supply chain impact of the proposed regulation on mid-sized logistics companies."

The response appeared almost instantly.

A structured summary.
Key regulatory provisions.
Operational implications.
A concise explanation of potential cost increases.

The reasoning looked sound. The explanation was clear. The structure felt complete.

The analyst read through it once, copied several points into a draft memo, and moved to the next task.

The work progressed efficiently.

What the analyst did not notice was that the first stage of thinking had quietly moved outside the mind.

Instead of constructing the structure of the problem internally, the analyst had received the structure from the system.

This shift can be described as **cognitive offloading**.

Earlier tools changed where information was stored. Artificial intelligence changes where ideas come from.

The distinction appears small at first glance, but it alters the structure of thinking in ways earlier tools did not.

Humans have always used external tools to extend cognitive ability.

Writing allowed information to exist outside memory. Maps reduced the need to memorize geography. Calculators eliminated the need for complex mental arithmetic. Each of these tools moved some cognitive work beyond the boundaries of the mind.

But those tools primarily assisted with storage and retrieval.

Artificial intelligence extends the process further.

It assists with idea generation.

And that difference changes how thinking begins.

For decades psychologists have studied how people distribute knowledge between internal memory and external systems.

In a widely cited experiment, psychologist Betsy Sparrow and her colleagues examined what happens when individuals believe information will remain accessible through a computer. Participants were asked to remember a series of facts. Some were told the information would be saved on a computer. Others were told it would be erased.

The results were striking.

When participants believed the information would remain digitally accessible, they were significantly less likely to remember the information itself. Instead, they remembered where the information could be found.

The brain adapted.

Rather than storing the knowledge internally, it stored a pathway for retrieving it.

Researchers later described this phenomenon informally as the **Google Effect**. It was not a failure of memory. It was an efficient strategy. When information can be retrieved instantly, remembering where it lives can be more useful than remembering the information itself.

A related idea had been described earlier by psychologist Daniel Wegner through the concept of **transactive memory**.

In collaborative environments people rarely attempt to memorize everything themselves. Instead they distribute knowledge across the group. One person understands financial details. Another knows legal regulations. Another retains technical expertise.

Members of the group remember who knows what.

Knowledge becomes a shared system rather than an individual one.

Artificial intelligence increasingly occupies a similar role.

Instead of remembering certain forms of knowledge internally, individuals remember that the system can generate it when needed.

At first glance this adaptation appears harmless. In many situations it is.

But a subtle shift occurs when offloading expands beyond information storage and into idea generation.

When a system provides the first structure of explanation, the mind encounters a finished argument rather than constructing one.

And constructing explanations is one of the primary ways expertise develops.

Learning does not occur simply by reading correct answers.

It occurs through the effort required to produce them.

Psychologist John Sweller's research on cognitive load theory helps explain why.

Sweller showed that expertise develops through the construction of internal knowledge structures known as **schemas**.

Schemas are mental frameworks that allow experienced professionals to interpret complex situations quickly and accurately. An experienced engineer looking at a design problem does not evaluate each variable from scratch. The relevant relationships appear almost immediately because the underlying structures already exist in memory.

But those structures do not appear automatically.

They form when individuals actively organize information, test interpretations, and connect ideas. This process requires effort.

Sweller described the effort responsible for building these internal structures as **germane cognitive load**. Germane load is the mental work that produces understanding.

When someone struggles to explain a concept, reorganizes information into a coherent model, or works through a difficult problem step by step, the brain is constructing schemas.

Artificial intelligence can unintentionally reduce this effort.

When the system produces a complete explanation instantly, the stage where the user would normally struggle to construct the explanation may never occur.

The explanation still appears useful.

But the mind engages with it differently.

Instead of building the structure, it evaluates one that already exists.

The difference becomes clearer when you compare two situations.

Imagine a junior analyst attempting to explain a regulatory conflict. Without assistance the analyst might begin by mapping the relevant statutes on paper. Several interpretations would be tested. Some would fail. Others would partially explain the interaction but leave inconsistencies unresolved.

Eventually a workable explanation would emerge.

The process would feel inefficient.

But during that process the analyst would construct a mental model of how the statutes interact.

Now imagine the same analyst receiving a finished explanation from an AI system.

The explanation may be correct.

But the mental process changes.

The analyst reads the argument, evaluates whether it sounds plausible, and perhaps improves its clarity.

The reasoning is reviewed.

But it was not constructed.

In the first case, the mind builds a schema.

In the second case, the mind encounters one.

Both workflows produce answers.

Only one consistently builds expertise.

A similar conclusion appears in research on expertise development.

Psychologist Anders Ericsson demonstrated that expert performance emerges through **deliberate practice**, sustained engagement with tasks that stretch the boundaries of current ability. Deliberate practice requires effort. It requires struggling with problems that do not yet have obvious solutions.

When individuals attempt to construct explanations themselves, make mistakes, and refine their reasoning, their internal models gradually become more sophisticated.

This process depends on generative effort.

Convenience can interrupt it.

If solutions consistently appear before individuals attempt to produce their own reasoning, the opportunities for that effort decrease.

Work still gets completed.

But the internal structures that normally develop through problem solving may form more slowly.

This shift is easy to miss because productivity remains high.

The analyst in the earlier example still produced a clear memo. The system's explanation may even have been correct. The work was completed faster than before.

Nothing appears broken.

Yet the role of the human mind during the task has changed.

Instead of generating the framework of the analysis, the analyst evaluated a framework produced externally.

Evaluation remains valuable.

But generation plays a different role in cognitive development.

Professionals who repeatedly construct frameworks themselves develop an internal architecture for interpreting complex problems. They learn which variables matter, which assumptions tend to fail, and which patterns signal deeper structural issues.

Those patterns become visible only after repeated attempts to construct explanations independently.

When external systems consistently supply the initial structure of reasoning, fewer opportunities for that generative work occur.

The difference is subtle in the moment.

It becomes visible over time.

Two professionals using the same systems may appear equally productive. Both produce clear reports. Both deliver persuasive presentations. Both rely on similar tools.

But their cognitive development may diverge.

One consistently begins thinking with the system.

The other begins thinking with the problem.

The first evaluates explanations.

The second constructs them.

Both workflows appear efficient.

But they build different kinds of expertise.

This distinction matters because the effects described in the previous chapter begin to compound.

Fluency makes generated explanations persuasive. Cognitive offloading moves the act of generation itself outside the mind.

Once both occur, the next change becomes almost inevitable.

The authority behind the reasoning begins to shift.

And that shift is the subject of the next chapter.

Exercises

Exercise 1: Generation Before Retrieval

Choose a professional topic you work with regularly.

Before consulting any AI system or external source, write a rough explanation of the central problem in the area, the key variables, the relationships between them, the most common point of confusion.

Give yourself ten minutes. The explanation does not need to be polished or complete.

Then ask an AI system to explain the same topic.

Compare the two.

The goal is not to evaluate quality. The goal is to notice what the act of generating produced. What connections formed? Which gaps became visible, and which questions arose, that reviewing the system's explanation would not have surfaced?

Exercise 2: Catch the Starting Point

During your next three work tasks involving AI assistance, note one thing before you begin each task: did you open the system first, or did you begin with the problem itself?

If you opened the system first, write down what structure the system provided. Then ask: would I have generated a similar structure on my own if given five minutes?

If you began with the problem, note how long you stayed with it before consulting the system. Was it long enough to construct a preliminary framework, or did you move to the system before any structure formed?

Most people discover the shift to the system happens earlier than they expected, often before any internal structure has begun to form.

Exercise 3: The Two-Analyst Test

Think of a problem you solved recently with significant AI assistance.

Reconstruct how the work unfolded: did you generate the initial framing of the problem, or did the system generate it? Who determined which variables mattered? Who structured the explanation?

Now imagine a colleague who approached the same problem type fifty times without AI assistance, constructing each explanation independently. What internal models would they have built that your workflow did not require you to build?

The goal is not to evaluate whether the AI-assisted approach produced worse work. The goal is to identify which cognitive development the workflow did and did not require.

CHAPTER 3

AUTHORITY TRANSFER

WHEN APPROVAL REPLACES ORIGINATION

The analyst expected the number to look wrong.

He had asked the system to estimate the market impact of a regulatory change on a mid-sized logistics company. The prompt was straightforward. The output arrived in seconds: a structured explanation followed by a revenue projection.

The number was precise.

$87 million in projected exposure.

He leaned back and read the explanation again. It sounded coherent. The reasoning flowed smoothly from one step to the next. The assumptions seemed reasonable.

Still, something about the estimate felt high.

For a moment he considered rebuilding the calculation himself. The underlying inputs were available. With enough time he could reconstruct the model step by step and see whether the system's reasoning held up.

But the report was due in an hour.

So he did what most professionals now do in this situation.

He scanned the explanation once more, made a minor wording adjustment to the conclusion, and sent the analysis forward.

The number might have been right.

It might have been wrong.

What mattered was that he never actually took ownership of it.

He had not produced the reasoning.

He had evaluated it.

That quiet shift is the subject of this chapter.

Earlier we saw why fluent explanations are so easy to trust. When reasoning arrives smooth and confident, scrutiny naturally relaxes. And once generative tools begin producing the first version of an idea, the mind often accepts the convenience of starting there.

Authority transfer is what happens next.

When fluent reasoning originates outside the mind, the user's role changes. The person who once constructed the argument becomes the person who decides whether the argument seems acceptable.

The shift appears small.

But it alters the structure of judgment itself.

Evaluating reasoning is not the same cognitive activity as producing it.

Psychologists who study decision-making have long observed that people are reasonably good at critiquing ideas when weaknesses are obvious. What we are much worse at is detecting hidden weaknesses inside arguments that appear coherent.

When you build an argument yourself, every step that produced the conclusion is visible. When the reasoning arrives finished, you encounter only the final structure.

The difference resembles the difference between building a bridge and inspecting one.

The builder understands every joint and support because they assembled it.

The inspector sees the completed structure and looks for weaknesses that might not be immediately visible.

Inspection requires a different mindset.

It also requires time.

When the role shifts from builder to inspector but the pace of work remains the same, authority quietly moves toward the system that produced the structure in the first place.

This is authority transfer.

The pattern becomes visible in the way professionals now interact with AI-generated analysis.

A system produces a structured explanation.

The user reads it quickly, scanning for obvious mistakes.

If none appear, the explanation is accepted.

This feels like verification.

In practice it is often a very shallow form of it.

True evaluation requires questions that rarely arise when reasoning arrives fully formed.

Are the assumptions complete?

Are there alternative explanations?

What information would falsify this conclusion?

Those questions normally appear while reasoning is being constructed.

When the explanation appears first, they appear less often.

Fluency was already reducing scrutiny.

Authority transfer adds a second layer to the problem: the reasoning being accepted may never have been yours to begin with.

Over time the structure of professional judgment shifts subtly.

Instead of asking

What conclusion should I reach?

The user increasingly asks

Does this conclusion seem reasonable?

The first question generates reasoning.

The second evaluates it.

That difference seems minor in isolation. But repeated across hundreds of decisions, it changes where intellectual authority resides.

The system proposes.

The human approves.

And once approval becomes the primary role, the system quietly becomes the originator of analysis.

This transition rarely feels dramatic.

It emerges gradually from the convenience of beginning with a fluent proposal instead of a blank page.

But convenience reshapes habits.

And habits reshape judgment.

Once this shift becomes visible, it suggests a simple diagnostic question.

Did I originate this reasoning, or did I approve it?

The answer reveals where authority currently sits in the workflow.

If the reasoning began inside your own thinking, the system likely played a supporting role.

If the reasoning appeared fully formed and your task was simply to decide whether it seemed acceptable, authority may already have shifted outward.

The goal is not to eliminate generative assistance.

The goal is to remain conscious of when judgment has moved from creation to evaluation.

Because evaluation requires a different discipline of mind.

And that discipline can be practiced deliberately.

Exercises

Exercise 1: The Origin Test

During your next task involving AI assistance, pause before submitting the final result and ask one question:

Did the core reasoning originate from me, or from the system?

If the system produced the initial explanation or structure, note how much of it remained unchanged.

Repeating this question regularly makes authority transfer visible.

Exercise 2: Reverse the Sequence

Choose a problem you would normally give directly to an AI system.

Before consulting it, write a brief explanation of the answer in two or three sentences.

Then ask the system for its version.

Compare the two explanations.

Evaluating a system's reasoning feels very different when you already possess a structure of your own.

Exercise 3: The Skeptic's Question

When reading a fluent AI-generated explanation, ask one additional question before accepting it:

What evidence would prove this conclusion wrong?

Write down at least one possible answer.

This small habit shifts the mind from passive approval to active scrutiny.

Fluent tools make reasoning appear quickly.

Generative systems make reasoning appear fully formed.

Together, they create an environment where the easiest role for the user is to approve arguments rather than construct them.

But authority in judgment has always belonged to the person

willing to do the harder work of asking whether the argument itself should exist.

When that work disappears, authority follows it.

And when authority moves quietly enough, it can transfer long before anyone notices the moment it left.

CHAPTER 4

REACTIVE MINDS

WHEN GENERATION BECOMES REACTION

At 9:12 a.m., the consultant opened the first prompt window.

A client presentation was due the next morning. The topic was operational efficiency in mid-sized manufacturing firms. She had a rough outline in mind, but before writing anything she asked the system for possible frameworks.

Three appeared instantly.

She skimmed them, selected one, and asked a follow-up prompt to expand the second section.

While waiting for the response she opened another window and asked for recent statistics on manufacturing automation.

Two minutes later she copied a paragraph from the output into the slide deck.

Then another idea occurred to her. She opened a third prompt window and asked for examples of companies that had implemented similar changes.

The response arrived quickly. She scanned it and pasted two examples into the presentation.

At 9:21 she moved to the next slide.

The work felt efficient.

Ideas were appearing quickly. Information was arriving instantly. The presentation was growing slide by slide.

But something subtle had happened during those nine minutes.

She had never once stayed with the problem long enough to develop a thought of her own.

Every step had been a response to the last output the system produced.

Her thinking had become reactive.

That shift, from generative thinking to reactive thinking, is the subject of this chapter.

Decades ago, economist and cognitive scientist Herbert Simon made an observation that feels increasingly prophetic:

A wealth of information creates a poverty of attention.

His insight was simple. Information does not merely inform us. It consumes attention.

Every piece of information requires the mind to process it, evaluate it, or decide whether it matters.

The more information appears, the more attention is spent reacting to it.

And attention spent reacting cannot be spent generating.

To understand the shift Simon anticipated, it helps to distinguish between two cognitive modes.

Generative thinking constructs ideas.

Reactive thinking responds to inputs.

When someone sits with a problem long enough to form a hypothesis, connect concepts, or build a new explanation, they are thinking generatively. The mind is producing structure.

Reactive thinking is different.

A stimulus appears, a message, a document, a suggestion, a prompt response, and the mind reacts to it. It edits, selects, approves, or rejects.

Both modes are necessary.

But they produce very different kinds of work.

Generative thinking creates arguments.

Reactive thinking modifies them.

The difference becomes clearer when you look at the rhythm of modern knowledge work.

Before generative AI systems existed, professionals still switched tasks frequently. Email arrived. Messages appeared. Documents required edits.

But extended periods of thinking still occurred between those interruptions.

The analyst worked through a model.

The writer struggled with a paragraph.

The strategist mapped possible scenarios.

Generative systems change that rhythm.

Because they produce new inputs continuously.

Every response suggests another prompt.

Every answer invites another question.

The workflow becomes a chain of reactions.

Instead of building ideas from within, the mind responds to outputs appearing from outside.

Reactive thinking gradually replaces generative thinking.

And because the outputs are useful, the shift feels productive.

This is where Simon's observation becomes more than a warning about distraction.

A wealth of information creates a poverty of attention.

Generative AI multiplies the wealth.

But it also multiplies the demands on attention.

These systems do not simply deliver information.

They produce structured explanations, outlines, strategies, and arguments.

Each one invites evaluation.

Each one becomes the stimulus for another reaction.

The mind enters a continuous loop of response.

And every response consumes attention that might otherwise have been spent generating an idea.

When thinking becomes reactive, the mechanisms described in earlier chapters intensify.

Fluent explanations become harder to question.

A reactive mind reads quickly, scanning for usefulness rather than examining structure.

Cognitive offloading increases.

When attention is fragmented, sustaining the effort required for internal idea generation becomes harder, so external systems become the natural starting point for thinking.

Authority transfer accelerates.

If the mind spends most of its time evaluating suggestions rather than constructing arguments, the system producing those suggestions quietly becomes the originator of analysis.

Attention fragmentation does not create these effects.

It accelerates them.

Cal Newport anticipated part of this problem in his work on deep work.

He argued that sustained attention is necessary for cognitively demanding tasks and warned that constant interruptions erode the ability to think deeply.

But the environment he described was primarily one of distraction.

Email notifications.

Social media.

Open-office interruptions.

Generative AI introduces a different challenge.

The interruption is not frivolous.

The system's responses are useful.

They contain ideas worth reading.

And that usefulness makes the attention drain harder to recognize.

Distraction pulls attention away from work.

Reactive cognition fragments the work itself.

Once this pattern becomes visible, a simple question reveals where thinking began.

Did the work start with the problem?

Or with a prompt?

When thinking begins with a prompt, the mind enters reactive mode immediately.

When thinking begins with the problem itself, generative thinking has space to emerge.

The order determines the mode.

Most professionals working with AI now start in reactive mode without realizing it.

A question appears.

A prompt is written.

An answer appears.

The mind reacts.

Another prompt follows.

The cycle continues.

Work progresses quickly.

But the quiet space where ideas normally form inside the mind becomes smaller.

Simon's insight echoes again here.

A wealth of information creates a poverty of attention.

And when attention becomes scarce, generative thinking is the first casualty.

Exercises

Exercise 1: The Five-Minute Rule

Choose a problem you would normally give directly to an AI system.

Before opening a prompt window, spend five minutes writing down your own thoughts about the problem.

They do not need to be polished or complete.

The purpose is simply to allow generative thinking to begin before reactive interaction starts.

Exercise 2: Catch the Shift

During your next task involving AI assistance, notice the precise moment your thinking changes.

At what point did you stop developing the idea internally and begin reacting to outputs from the system?

Write that moment down.

Repeat this observation during several tasks.

Most people discover that the shift happens much earlier than they expected.

Exercise 3: Attention Mapping

At the end of a task completed with AI assistance, ask yourself two questions:

How much time did I spend reacting to outputs?

How much time did I spend developing ideas internally?

Write the estimates down.

Most people are surprised by how little generative time remains once reactive workflows take hold.

A wealth of information creates a poverty of attention.

Simon's observation becomes even more relevant in a world where information arrives not only as facts, but as fluent reasoning.

The mind reacts.

The system produces more.

The cycle continues.

And in that cycle, something subtle begins to disappear.

The quiet time in which ideas form inside the mind.

Without that space, thinking does not stop.

But it becomes reactive rather than generative.

And when thinking becomes reactive, the forces described in the previous chapters strengthen all at once.

Fluency becomes harder to question.

Ideas originate elsewhere.

Authority follows them there.

Not suddenly.

But prompt by prompt, response by response, until the mind forgets what it felt like to begin with a blank page.

CHAPTER 5

THE STRAIN

THE COST OF CONVENIENCE

Eight months into using AI in her legal practice, the work felt effortless.

Drafting motions that once required hours now took minutes. The system produced structured arguments instantly. It identified precedents, synthesized statutes, and suggested lines of reasoning before she finished describing the problem. The work still required review, but the structure arrived ready-made.

At first, the efficiency felt liberating. The system removed the tedious parts of the job. Time once spent assembling arguments could now be spent refining them.

But the pattern repeated often enough that the process began to shift. Instead of building arguments from the ground up, she began starting with the system's output. The role of thinking subtly changed. Instead of generating structure, she evaluated structure.

The work still felt intellectual. It still required judgment. But the strain that once accompanied difficult thinking appeared less frequently.

Then a case arrived that didn't fit the pattern.

Two statutes appeared to conflict in a way the system interpreted as independent regulatory authority. The AI produced a clear, fluent explanation of why both could apply simultaneously. The reasoning looked sound at first glance.

But something felt wrong.

The statutes overlapped in a narrow jurisdictional condition the system had treated as unrelated. The conflict wasn't explicit. It required reconstructing the legislative structure behind both provisions.

She recognized the error immediately.

What followed was unexpected.

Correcting the system required rebuilding the argument from scratch. That meant reconstructing the legislative logic that connected the statutes in the first place.

And that was harder than it should have been.

Not because the reasoning was impossible, but because the strain required to generate the structure independently felt unfamiliar.

The system had been producing the scaffolding of arguments for months. Her role had been to evaluate and refine it.

Now the scaffolding was gone.

The strain that once accompanied difficult thinking had quietly disappeared from the workflow.

And with it, some of the internal structures that made difficult thinking possible.

Productive Effort versus Generative Effort

The difference between those two experiences can be described with a simple distinction.

Productive effort finishes tasks.
Generative effort builds capability.

Productive effort is the effort required to complete work.

Writing a brief.
Drafting a report.
Summarizing a case.
Producing an answer.

AI systems excel at reducing productive effort. That is their primary advantage.

Generative effort is different.

Generative effort occurs when the mind must construct understanding rather than apply it. It is the strain involved in organizing ideas, testing interpretations, and discovering relationships between concepts.

Generative effort does not merely finish work. It builds the internal structures that make future thinking possible.

The difficulty is that the two forms of effort often appear identical during the moment of work. Both feel like thinking. Both feel like progress.

But over time they produce very different outcomes.

Productive effort produces output.

Generative effort produces expertise.

When AI systems remove productive effort, the gain is efficiency.

When they remove generative effort, the cost is capability.

The danger is not that AI completes work quickly.

The danger is when the work it replaces was the very process through which expertise was forming.

What Deliberate Practice Requires

The psychologist Anders Ericsson spent decades studying how expertise develops across fields ranging from music and athletics to medicine and law.

His research identified a consistent pattern.

Expertise does not emerge from repetition alone. It develops through **deliberate practice**, a specific form of effort characterized by three conditions:

1. The task occurs at the **edge of current capability**.

2. **Feedback reveals failure or error**.

3. The environment allows the **possibility of failure**.

Each condition forces the mind to construct new internal structures.

Working at the edge of capability prevents reliance on automatic routines. Feedback reveals gaps in understanding. The possibility of failure creates pressure to refine the internal model of the problem.

This combination is what produces expertise.

It is also precisely what AI systems tend to eliminate.

AI lowers the difficulty of tasks by providing immediate structure.

It supplies answers before feedback becomes necessary.

And by offering fluent outputs, it reduces the frequency with which failure becomes visible.

The result is efficient work.

But the conditions that produce expertise begin to disappear.

The edge of capability becomes easier to bypass.

Feedback becomes less informative.

Failure becomes less frequent.

The mind still completes tasks.

But it encounters fewer situations where it must build new internal structures to solve them.

Desirable Difficulties

A similar conclusion appears in the work of cognitive psychologist Robert Bjork.

Bjork introduced the concept of **desirable difficulties**, the counterintuitive finding that learning is often strengthened by conditions that make the process feel harder.

Spacing practice over time improves retention even though it feels less efficient than cramming.

Testing memory strengthens understanding more than rereading material, even though it feels more difficult.

Interleaving topics improves transfer of knowledge, even though it initially slows performance.

In each case, the condition that improves learning is the condition that increases strain.

The difficulty forces the mind to reconstruct knowledge rather than recognize it passively.

This insight carries an uncomfortable implication.

The efficiency that makes AI assistance feel productive may simultaneously weaken the conditions under which learning occurs.

The system removes friction.

But friction is often the signal that generative effort is happening.

When work becomes effortless, it can become informationally thin.

Output continues.

But internal structure stops forming.

The Fifth Effect

Across the previous chapters, we examined four ways AI systems reshape thinking.

Fluency can distort credibility.

Generation can replace retrieval.

Authority can transfer from the thinker to the system.

Attention can fragment into reactive interaction.

Each of these effects alters the structure of thinking in the moment.

But over time a deeper effect emerges.

Expertise begins to atrophy.

Not because professionals stop working.

Not because intelligence declines.

But because the strain that once constructed internal models of complex problems appears less frequently.

The professional who never struggles eventually encounters a situation where struggle is required.

And discovers the structures that once supported it are weaker than they used to be.

Expertise does not disappear all at once.

It erodes quietly.

The absence of strain feels like efficiency.

Until the moment when independent reasoning becomes necessary again.

The Hidden Cost of Convenience

In many professional environments, the goal is simple: reduce the time required to produce competent work.

A contract draft that once required an afternoon now arrives in minutes.
A technical memo appears fully structured before the first outline is written.
A literature review assembles citations before the researcher has finished defining the question.

These developments genuinely improve productivity.

But they also change how capability develops.

Convenience removes the effort involved in assembling arguments, organizing information, and constructing explanations.

Those tasks are exactly where generative effort used to occur.

When productive effort disappears, efficiency increases.

When generative effort disappears, expertise stops growing.

The distinction matters most over time.

A professional who regularly encounters generative strain continues building internal structures for reasoning under uncertainty.

A professional who rarely experiences that strain becomes increasingly dependent on external systems to generate those structures.

Both may appear equally capable while the system is available.

The difference becomes visible only when the system cannot provide the answer.

That moment reveals whether capability was built internally.

Or borrowed.

Exercises

Exercise 1: Reconstruction

Choose a recent decision or piece of work where you relied heavily on AI assistance.

Without consulting the original output, reconstruct the reasoning behind the solution entirely from memory.

Write the argument or logic step by step.

Notice where the reasoning flows easily and where gaps appear.

Those gaps mark places where generative effort was replaced by externally produced structure.

Exercise 2: Remove the Scaffold

During your next AI-assisted task, ask the system for **information only**, not conclusions.

Collect the relevant data, sources, or background material.

Then construct the reasoning and final answer independently.

Compare your reasoning to the system's afterward.

The goal is not to outperform the system but to experience the generative strain of building structure yourself.

Exercise 3: Productive versus Generative

At the end of a workday, identify three moments where you solved a problem.

For each moment, ask:

Did this effort **complete a task**, or did it **build a new internal understanding**?

Over time, aim to maintain regular exposure to problems that require generative effort.

Capability grows where strain occurs.

The Discipline of Strain

The convenience of AI will continue to improve.

Systems will become faster, more fluent, and more capable of generating structured answers.

This progress will remove increasing amounts of productive effort from professional work.

But the removal of effort creates a choice.

The mind can accept the convenience entirely and allow generative strain to disappear from the workflow.

Or it can preserve moments of difficulty intentionally.

Not because difficulty is virtuous.

But because difficulty builds the structures that make judgment possible.

Expertise does not form in moments of ease.

It forms in moments when thinking must generate its own structure.

The question is not whether AI will reduce effort.

It will.

The question is whether the effort that disappears was the kind that only finished work.

Or the kind that once built the ability to think.

CHAPTER 6

AUTHORSHIP

WHEN EDITING REPLACES AUTHORING

The presentation had gone smoothly.

The analyst walked through the slides with confidence. The conclusions were clear. The reasoning appeared structured and persuasive. The room nodded along as the argument unfolded.

Then a colleague asked a simple question.

"Can you explain why the second assumption follows from the first?"

It wasn't a hostile question. It was exactly the kind of clarification people routinely ask when reviewing analysis.

The analyst paused.

The slide showed the reasoning clearly enough. The logic appeared structured and coherent. But explaining it required reconstructing the argument step by step.

And that proved unexpectedly difficult.

Not because the reasoning was wrong.

But because the reasoning had never been constructed by the analyst in the first place.

It had been generated by the system.

The analyst had reviewed it carefully. Adjusted wording.

Rearranged sections. Improved clarity. Approved the final version.

But the internal structure of the argument had arrived already assembled.

In the moment of questioning, the difference became visible.

The analyst had presented the reasoning.

But the reasoning was not entirely theirs.

Authoring versus Editing

This difference can be described with a simple distinction.

Authoring generates reasoning.
Editing modifies reasoning.

Authoring occurs when the structure of an argument originates within the thinker's own process of reasoning. The thinker determines the relationships between ideas, constructs the explanation, and builds the logic that connects premises to conclusions.

Editing occurs when the structure already exists.

The editor evaluates it, improves it, reshapes it, or corrects it.

Both activities are legitimate. Both require intelligence.

But they produce different cognitive outcomes.

Authoring builds the internal structures that allow reasoning to occur.

Editing evaluates structures that have already been created.

The difference matters because modern AI workflows make it

increasingly possible to edit endlessly without ever authoring.

The reasoning arrives first.

The user participates afterward.

Recognition versus Understanding

This dynamic is strengthened by a well-known cognitive phenomenon identified by psychologists Leonid Rozenblit and Frank Keil.

They called it the **illusion of explanatory depth**.

When people encounter explanations they can recognize and follow, they often believe they understand them far more deeply than they actually do.

The difference becomes visible only when they are asked to explain the mechanism themselves.

People who confidently claim they understand how a toilet works, for example, often struggle to explain the flushing mechanism step by step.

Recognition creates the feeling of understanding.

But recognition and understanding are different cognitive states.

Recognition means the explanation appears plausible.

Understanding means the explanation can be reconstructed independently.

AI-generated reasoning amplifies this illusion dramatically.

When a system produces a coherent argument and the user reads it carefully, the reasoning becomes familiar.

Familiarity feels like comprehension.

But when someone asks the user to explain the argument from first principles, the difference between recognition and understanding appears immediately.

The explanation cannot be reconstructed because the internal structure was never built.

It was only observed.

The Procedural Shift

Across the previous chapters, we examined how AI systems alter the structure of thinking.

Fluent output can distort credibility.

Generated knowledge can replace retrieval.

Authority can transfer from thinker to system.

Attention can become reactive rather than generative.

Generative effort can disappear from the workflow.

Chapter 6 examines the professional consequence of those shifts.

Authorship changes.

The traditional process of professional reasoning follows a recognizable sequence.

1. Define the problem

2. Retrieve relevant knowledge

3. Construct the reasoning structure

4. Test the argument

5. Refine the conclusion

In an AI-assisted workflow, the sequence often changes.

1. Describe the problem

2. Receive generated reasoning

3. Evaluate plausibility

4. Modify wording or structure

5. Approve the result

The most important step in the original process, constructing the reasoning structure, quietly disappears.

The thinker still participates.

But participation is no longer the same as authorship.

The Responsibility Problem

This shift introduces a subtle but significant professional risk.

Professionals are not judged only by the quality of their conclusions.

They are judged by their ability to **defend the reasoning behind them**.

Lawyers must defend arguments.
Engineers must justify designs.
Scientists must explain mechanisms.
Leaders must account for decisions.

When reasoning originates internally, defending it is natural.

The thinker knows how the structure was built because they built it.

When reasoning arrives externally, the relationship changes.

The professional may still review the reasoning carefully.

They may agree with it.

They may even improve it.

But agreement is not the same as authorship.

Consider a common moment in technical work.

An engineer presents a system design during a review. The design document is well organized and persuasive. But during the discussion, a senior colleague asks a simple question.

"Why does the load calculation assume that constraint?"

The engineer looks back at the document.

The assumption is there.

The explanation looks correct.

But reconstructing the reasoning behind it requires tracing the entire calculation chain.

The engineer approved the design.

But the design logic itself came from the system.

And in that moment, the distinction between reviewing reasoning and owning it becomes visible.

The question is no longer whether the design appears correct.

The question is whether the engineer can **reconstruct the reasoning independently**.

That ability is the foundation of intellectual responsibility.

When the Effects Converge

The previous chapters described five ways AI systems reshape individual cognition.

Individually, each effect is subtle.

Fluency can make generated explanations appear trustworthy.

Externally generated knowledge can replace internal retrieval.

Authority can shift from thinker to system.

Attention can become reactive.

Generative effort can disappear from the workflow.

In isolation, none of these changes necessarily breaks professional reasoning.

But the authoring-versus-editing shift is where all five effects become visible at once.

Fluent reasoning arrives already constructed.

The knowledge behind it was generated rather than retrieved.

The system's structure becomes the authority the user evaluates.

The workflow becomes reactive rather than generative.

And the strain that once built internal reasoning structures is absent.

The result is a professional who can recognize strong reasoning, improve it, and present it convincingly.

But when asked to rebuild the reasoning independently, the structure is missing.

Authorship is where the cumulative effects of AI-assisted thinking finally surface.

Exercises

Exercise 1: The Feynman Test

Choose a recent argument, analysis, or explanation you produced with AI assistance.

Without consulting the original output, attempt to explain the reasoning step by step as if teaching it to someone unfamiliar with the topic.

Use simple language.

Avoid referencing the system's phrasing.

If the explanation becomes difficult to reconstruct, you have discovered where authorship was external rather than internal.

Exercise 2: Reverse the Process

During your next AI-assisted task, reverse the typical workflow.

First construct the argument structure yourself.

Write the reasoning in outline form.

Then use AI only to critique or refine that structure.

This preserves the authoring stage while still benefiting from AI assistance.

Exercise 3: Ownership Check

Before presenting or submitting AI-assisted work, ask yourself one question:

Could I defend every step of this reasoning without the system present?

If the answer is uncertain, the reasoning may have been edited rather than authored.

The Moment of Exposure

Editing can produce excellent work.

It can improve clarity, strengthen arguments, and refine analysis.

But the mind that edits reasoning is not the same mind that builds it.

One evaluates.

The other constructs.

Construction is where understanding forms.

Which is why the difference often appears only in moments of pressure.

A colleague asks a simple question.

A client asks for an explanation.

A regulator asks for justification.

And suddenly the reasoning must be reconstructed rather than recognized.

In that moment, the distinction between editing and authoring becomes unmistakable.

The professional who built the reasoning can walk through it step by step.

The professional who only reviewed it looks back at the slide.

And realizes the answer should be obvious.

But the structure that produced it was never theirs to begin with.

CHAPTER 7

THE SIGNAL PROBLEM

PROTECTING ATTENTION FROM USEFUL INTERRUPTION

The researcher sat with the problem longer than usual.

The AI interface was open on the screen. The cursor blinked in the prompt box, waiting.

Normally the next step would have been obvious. Describe the question, ask for possible explanations, review the generated reasoning.

But this time the researcher waited.

For the first ten minutes, nothing happened. The mind drifted between fragments of half-formed ideas and possible interpretations. The absence of immediate structure felt inefficient.

Twenty minutes passed.

Slowly, something began to emerge. A possible relationship between two variables that hadn't seemed connected before. Then a second idea that complicated the first. The researcher sketched a rough diagram on paper, testing whether the relationship held under slightly different assumptions.

The diagram didn't work at first.

Then a small modification made the structure clearer.

Forty minutes later, the researcher finally turned to the system.

The prompt looked different this time.

Instead of asking the system to generate possible explanations, the researcher described a specific model and asked the system to challenge it.

The response that followed was more useful than usual.

Not because the system was better.

But because the thinking that preceded the interaction had created something for the system to test.

The AI interaction had become a critique of internally generated reasoning rather than a replacement for it.

The difference began with forty minutes of resistance.

Noise versus Signal

Protecting attention used to mean something relatively simple.

Remove distractions.

Silence notifications. Close unnecessary tabs. Step away from email. Create uninterrupted time for concentrated thinking.

This was the problem described by Cal Newport in *Deep Work*. The modern environment fragments attention through constant interruptions, and protecting attention requires reducing those interruptions.

In a pre-AI environment, this discipline worked well because most interruptions were trivial.

They were noise.

A notification rarely contained information more valuable than the thinking it interrupted.

AI systems introduce a different kind of interruption.

When the mind reaches for the system, the result is not trivial information.

The result is often useful reasoning.

Suggestions. Counterarguments. Structured explanations.

These interruptions are not noise.

They are signal.

Which creates a fundamentally different discipline problem.

The difficulty is no longer protecting attention from meaningless distraction.

The difficulty is protecting attention from **useful interruption**.

And useful interruptions are much harder to resist.

Why Signal Is Harder to Resist

Resisting noise requires discipline.

Resisting signal requires judgment.

Noise can be ignored because it adds little value.

Signal is difficult to ignore precisely because it might contain something valuable.

During those forty minutes, the researcher knew exactly what the system could offer.

It could propose possible explanations for the relationship between the variables.

It could generate models the researcher had not yet considered.

It could produce a diagram that looked more complete than the uncertain sketch forming on paper.

That possibility makes the interruption difficult to resist.

Why struggle through fragments and half-formed ideas when the system can instantly produce something coherent?

But those fragments were the beginning of generative thinking.

The mind had to struggle through incomplete models before a workable structure appeared.

If the system had entered the process earlier, the structure would have arrived externally.

The researcher would have evaluated it rather than constructed it.

Attention as the Precondition of Authorship

The previous chapter showed that authorship requires constructing the internal structure of reasoning.

But construction requires something even more basic.

Time.

Specifically, sustained attention.

Without uninterrupted cognitive space, the mind cannot explore possibilities deeply enough to construct new frameworks of understanding.

It can react.

It can evaluate.

It can edit.

But it cannot author.

Which means attention discipline is not merely a productivity technique.

It is the condition that makes authorship possible.

If attention constantly moves from prompt to response, the mind rarely occupies the generative state long enough to build the reasoning structures that later chapters depend on.

The authoring-versus-editing distinction therefore begins with a simpler choice.

When does the AI interaction begin?

At the beginning of thinking.

Or after thinking has begun.

The Discipline of Delay

In a world filled with noise, attention discipline meant removing distractions.

In a world filled with signal, attention discipline means delaying assistance.

The discipline is simple to describe but difficult to practice.

Sit with the problem first.

Allow the mind to construct a preliminary model.

Let uncertainty exist long enough for ideas to form.

Only then consult the system.

This delay does not eliminate AI from the workflow.

It changes the role AI plays.

When the system appears after generative thinking has begun, the interaction becomes critique rather than creation.

The AI tests ideas rather than supplying them.

The professional who consults the system immediately risks becoming an editor of reasoning.

The professional who waits first remains its author.

Exercises

Exercise 1: The Delay Rule

Choose a problem you would normally approach with AI assistance.

Set a timer for thirty minutes.

During that time, work on the problem without consulting the system.

Sketch ideas, outline possible explanations, or construct a preliminary argument.

Only after the timer ends should you consult AI.

Compare the interaction with your typical workflow.

Notice whether the system responds to your reasoning rather than replacing it.

Exercise 2: Question Before Answer

Before consulting AI, write down the exact question you want the system to address.

Then write a preliminary answer yourself.

Even if the answer is incomplete.

The goal is not correctness.

The goal is to force the mind to generate structure before receiving one.

Exercise 3: Signal Awareness

During a workday, notice how often the impulse to consult AI arises.

Ask a simple question each time:

Is this interaction clarifying my thinking, or replacing it?

The difference determines whether the system is assisting authorship or quietly converting it into editing.

The Value of Resistance

In a distracted world, resisting interruption protected concentration.

In an AI-saturated world, resistance serves a different purpose.

It protects the generative stage of thinking.

The moments when ideas are still fragile.

When explanations are incomplete.

When understanding is still forming.

Those moments are uncomfortable.

They are slow.

They often feel inefficient.

But they are also where original reasoning begins.

If the system arrives too early, that stage disappears.

And the mind learns a different habit.

The habit of consulting before thinking.

Which is why the most important discipline in an AI-assisted environment is not avoiding the system.

It is resisting it just long enough for thinking to begin.

CHAPTER 8

INSTITUTIONAL CONVERGENCE

WHEN INDIVIDUAL DEFERENCE BECOMES COLLECTIVE UNIFORMITY

Chapter 3 described a subtle shift in individual cognition.

When AI systems generate fluent reasoning, professionals begin to trust the output before they have independently examined the underlying logic. The role of the thinker shifts from originator of analysis to evaluator of system-generated reasoning. Authority transfers.

That transfer occurs quietly. One professional at a time. One workflow at a time.

But when that same shift occurs across an entire organization, something different happens.

Authority does not simply move from a person to a system.

It concentrates.

When hundreds or thousands of professionals begin relying on the same analytical systems, their thinking begins to converge around the same frameworks, the same summaries, the same interpretations of evidence.

The result is not simply individual deference to AI.

It is institutional convergence.

Individual deference transfers authority.
Institutional convergence embeds it.

Authority no longer resides in the reasoning of any particular professional. It becomes embedded in the analytical infrastructure the organization relies upon.

And once authority becomes embedded in infrastructure, it becomes far harder to detect.

The Meeting

Consider a leadership team preparing for a strategic decision.

Each executive has spent the week analyzing the same question: whether the organization should enter a new market.

They arrive at the meeting prepared.

One emphasizes financial risk.

Another highlights operational constraints.

A third argues for a more aggressive strategy given projected growth.

The discussion appears healthy. Different voices raise different concerns. Multiple perspectives seem to be represented.

But beneath that apparent diversity lies something more subtle.

Each executive conducted their preparation with the assistance of the same AI system.

Each consulted similar prompts.

Each received similar research summaries, similar market analyses, similar frameworks for evaluating the opportunity.

The conclusions they reached are not identical.

But the frameworks organizing their thinking are.

The financial executive is using the system's recommended risk model.

The operations leader is relying on the system's summary of supply chain constraints.

The strategist is using the growth scenarios the system generated.

Three perspectives.

One analytical origin.

The team believes it is debating independently generated insights.

In reality, their reasoning has already converged before the meeting begins.

The decision that emerges will feel collectively owned.

But the structure of reasoning behind it originated somewhere else.

Informational Cascades Without Interaction

This convergence can be understood through research on informational cascades.

Legal scholar, behavioral economist, and policy theorist Cass Sunstein and his colleagues studied how people update their judgments when they observe others accepting a particular conclusion.

When individuals see others adopting a belief, they often shift their own views toward that belief, even if their private information points in a different direction.

Over time a cascade forms.

Once enough people adopt the same conclusion, the cascade sustains itself. Each additional participant reinforces the appearance that the conclusion must be correct.

Traditionally, informational cascades emerge through **social observation**.

People see what others believe and adjust their own beliefs accordingly.

AI systems create a different kind of cascade.

Here the convergence happens **before anyone interacts with anyone else**.

Each individual consults the same analytical system.

Each receives similar frameworks for understanding the problem.

Each forms conclusions shaped by those frameworks.

By the time the group meets, the cascade has already begun.

The convergence is not created by social pressure.

It is created by **analytical alignment**.

The participants believe their reasoning developed independently.

In reality, it was shaped by the same source.

Historical Precedent

Institutional convergence is not unique to AI.

Organizations have experienced similar dynamics whenever a single analytical framework achieved widespread dominance.

In the years leading up to the 2008 financial crisis, sophisticated risk models became standard tools across the financial industry.

Banks, rating agencies, and regulators all relied on similar frameworks for evaluating mortgage-backed securities.

Those frameworks appeared rigorous.

They were quantitative, mathematically elegant, and widely accepted.

But they shared the same blind spots.

They underestimated the possibility that housing markets across the country could decline simultaneously.

Because institutions relied on the same models, those blind spots became systemic.

When housing prices began falling, the failure was not isolated to one institution.

It was distributed across the entire system.

The convergence of analytical frameworks produced a convergence of error.

AI systems introduce the same risk, but with two important differences.

First, convergence can happen far more quickly.

Instead of gradually adopting a framework over years, organizations can converge around system-generated reasoning almost immediately.

Second, the frameworks are not always explicitly chosen.

They are generated dynamically by the system itself.

Professionals may not realize how similar their analytical foundations have become.

The convergence happens automatically.

Diversity of Voice, Uniformity of Thought

Institutional convergence is difficult to detect because it preserves the appearance of intellectual diversity.

People still speak in their own voices.

They raise different concerns.

They argue for different actions.

But the frameworks organizing those arguments may already be aligned.

The result is a paradox:

Diversity of voice.
Uniformity of thought.

Disagreement persists at the surface level.

But the deeper structure of reasoning has already converged.

When that happens, organizations can become highly confident in decisions that were produced by a narrow analytical lens.

The absence of visible disagreement reinforces the sense that the reasoning must be sound.

Yet the apparent consensus may simply reflect shared analytical origins.

The Consequence

When authority becomes embedded in analytical infrastructure, institutions gain speed and efficiency.

Decisions can be made more quickly. Analysis can be performed at greater scale.

But something else is lost.

Organizations begin to see the world through a smaller number of analytical lenses.

The system's assumptions become the organization's assumptions.

Its blind spots become institutional blind spots.

The risk is not that AI systems occasionally produce flawed reasoning.

The risk is that flawed reasoning becomes systematically reproduced across the institution.

When that happens, error is no longer individual.

It becomes structural.

From Minds to Institutions

The previous chapters examined what happens inside individual cognition when AI systems enter professional workflows.

Clarity begins to substitute for correctness.

Retrieval replaces generation.

Professionals shift from originating reasoning to approving system-generated outputs.

Attention becomes reactive rather than generative.

Effort migrates from thinking to editing.

These effects begin in individual minds.

But institutions are composed of individuals.

When those individual shifts occur simultaneously across an organization, their effects compound.

The result is institutional convergence.

Reasoning no longer originates primarily within the organization itself.

It flows through the analytical systems the organization depends upon.

Authority no longer resides primarily in professional judgment.

It becomes embedded in the infrastructure of reasoning.

Understanding that shift is essential.

Because the next question is not simply how individuals should think alongside AI systems.

The next question is how institutions should be designed so that the power of these systems strengthens human judgment rather than quietly replacing it.

Exercises

Exercise 1: Detecting Framework Influence

Think about the last time you used an AI system to analyze a complex issue.

After receiving the system's output, ask yourself two questions:

1. What conclusions did the system suggest?
2. What frameworks did it use to organize the problem?

Now ask a third question:

If I had analyzed this problem without the system, would I have used the same framework?

The goal of this exercise is not to reject system assistance.

It is to recognize when your reasoning has been shaped by analytical structures that originated elsewhere.

Awareness is the first defense against unconscious convergence.

Exercise 2: Recognizing Convergence in Groups

In your next team discussion or meeting, pay attention to how problems are framed.

Ask yourself:

- Are different participants organizing the problem using similar frameworks?
- Are similar assumptions appearing across multiple viewpoints?
- Do disagreements occur primarily about conclusions rather than the structure of reasoning?

If so, the group may already share an analytical foundation.

Recognizing convergence allows teams to deliberately introduce alternative perspectives before decisions are finalized.

Exercise 3: Preserving Cognitive Diversity

If you lead a team or organization, experiment with a simple structural change.

Before discussing a major decision, ask participants to prepare their analysis using **different sources of reasoning**.

Some may consult AI systems.

Others may rely on traditional research or independent modeling.

Still others may be asked to challenge prevailing assumptions directly.

The goal is not to reject algorithmic assistance.

It is to ensure that the organization maintains multiple analytical lenses rather than converging prematurely on a single one.

Cognitive diversity rarely survives by accident.

It must be designed.

CHAPTER 9

KNOWING VERSUS ACCESSING

WHAT THE MIND MUST STILL HOLD

A junior physician is reviewing a patient in the emergency department.

The patient presents with chest pain, mild shortness of breath, and fatigue. Nothing about the case appears immediately catastrophic, but the symptoms are ambiguous enough to require careful evaluation.

The physician opens an AI-assisted clinical system and enters the patient's symptoms.

Within seconds the system produces a structured summary of possible diagnoses, supporting research, and suggested next steps.

The information is accurate. The references are credible. The reasoning appears sound.

But something important has already happened.

Before the physician begins reasoning about the case, the system has already framed the problem.

The doctor's role becomes interpreting the system's possibilities rather than generating the diagnostic structure independently.

In most cases this may not matter.

But in situations where time is limited and uncertainty is high, the distinction becomes critical.

Because effective reasoning under pressure depends on knowledge that already exists inside the mind.

Not knowledge that must be retrieved before thinking can begin.

The Governing Distinction

The difference can be described with a simple distinction.

Knowing versus accessing.

Knowing means information is internalized.

It can be recalled instantly, connected to other knowledge, and used as the foundation for reasoning.

Accessing means the information exists somewhere retrievable.

A system can provide it when asked.

The difference appears small.

In practice, it determines how thinking begins.

Knowing allows reasoning to begin immediately.
Accessing requires retrieval before reasoning can begin.

When problems are simple and time is abundant, accessing information may be sufficient.

But complex reasoning rarely begins with a fully formed question.

It begins with fragments of understanding already present in the mind.

Those fragments interact, collide, and gradually organize the problem.

If the fragments are not present, reasoning cannot start.

It must wait.

The Hidden Cost of Unlimited Access

Modern knowledge systems create the impression that internal knowledge is becoming less necessary.

If information can be retrieved instantly, why memorize it?

Why internalize facts that can be accessed on demand?

This reasoning appears efficient.

But it misunderstands how human cognition works.

Reasoning is not simply the process of retrieving facts and assembling them into conclusions.

Reasoning emerges from **interactions between pieces of knowledge already stored in the mind**.

When those pieces are internalized, they can interact continuously.

Connections form spontaneously.

New questions arise during the act of thinking.

Insights emerge from the collision of ideas.

When knowledge exists only externally, this interaction cannot occur.

Before each step of reasoning, the thinker must pause and retrieve the next piece of information.

The cognitive process becomes fragmented.

The mind waits for the system.

Knowledge Compression

Expertise is often misunderstood as the accumulation of large quantities of information.

In reality, expertise is something more subtle.

Experts do not simply know more.

They know **more efficiently**.

Over time, large bodies of information become compressed into compact mental models.

A chess grandmaster does not analyze every possible move individually.

They recognize patterns that represent thousands of prior games.

An experienced firefighter entering a burning building does not consciously evaluate every variable.

Years of experience have compressed knowledge about smoke patterns, heat behavior, and structural risk into immediate recognition.

A seasoned investor reviewing financial statements does not calculate every ratio from first principles.

Certain signals immediately stand out because prior knowledge has been integrated into intuitive frameworks.

This process is called **knowledge compression**.

Complex information becomes organized into structures that can be applied quickly and flexibly.

Those structures allow reasoning to occur under pressure and uncertainty.

And they cannot be built through retrieval alone.

They require knowledge that has been internalized.

When Fragmented Knowledge Fails

The consequences of relying primarily on accessed information often remain invisible until conditions change.

Imagine the physician again in the emergency department.

A second patient arrives with a similar cluster of symptoms, but this time the situation evolves rapidly.

The patient's blood pressure begins dropping.

The room becomes tense. Nurses are preparing medications. A decision must be made immediately.

There is no time to consult a system.

An experienced clinician begins recognizing patterns instantly. The symptoms suggest a rare but dangerous condition that requires immediate intervention.

Another physician, equally intelligent but more dependent on external reference systems, hesitates.

The knowledge exists.

It is documented in textbooks and databases.

But it is not yet compressed into internal models that allow recognition under pressure.

Access is not the problem.

Time is.

Fragmented knowledge often fails not because information is unavailable, but because it is unavailable **inside the mind at the moment reasoning must occur**.

The Illusion of External Intelligence

AI systems amplify access to information in extraordinary ways.

They summarize research instantly.

They synthesize large datasets.

They generate plausible analytical frameworks within seconds.

These capabilities create a powerful illusion.

If intelligence can be accessed externally, perhaps it no longer needs to be developed internally.

But external intelligence cannot replace internal knowledge.

It can only supplement it.

External systems answer questions.
Internal knowledge determines **which questions to ask**.

External systems generate options.
Internal knowledge evaluates whether those options make sense.

External systems provide analysis.
Internal knowledge determines whether the analysis applies.

Without internal knowledge, professionals risk becoming highly efficient users of tools they cannot fully evaluate.

Returning to Generation

Chapter 2 introduced the importance of generation.

Understanding deepens when individuals produce reasoning themselves rather than simply retrieving answers.

Chapter 9 extends that idea.

Generation builds knowledge.

And knowledge, once internalized, becomes the compressed foundation that makes further reasoning possible.

Access can support that process.

But it cannot replace it.

The Final Diagnostic Piece

Across the previous chapters, we have examined how AI systems reshape professional cognition.

Fluent outputs alter how we judge correctness.

Retrieval replaces generation.

Authority shifts from thinker to system.

Attention becomes reactive rather than generative.

Effort moves from producing reasoning to editing it.

Organizations begin converging around shared analytical frameworks.

Chapter 9 completes the diagnostic picture.

When knowledge itself migrates outward into external systems, the mind risks losing the compressed structures that make reasoning possible under uncertainty.

Access expands.

Knowing contracts.

Professionals may appear more informed than ever before.

Yet the internal foundations required for independent judgment may quietly weaken.

The next chapter examines what happens when judgment must operate under uncertainty in exactly that environment.

Exercises

Exercise 1: Identifying Internal Knowledge

Choose a topic central to your profession.

Without consulting any external source, write down the key principles that guide your reasoning in that area.

Then compare your list with external references.

Which principles did you already know?

Which did you need to look up?

The goal of this exercise is to identify the core knowledge that currently exists inside your mind versus the knowledge you routinely access externally.

Exercise 2: Building Knowledge Compression

Identify one concept in your field that you typically access externally.

Spend time studying it until you can explain it clearly without consulting any reference.

Then compress it into a short explanation you could deliver under pressure.

The goal of this exercise is to practice transforming accessed information into internalized knowledge.

Exercise 3: Designing Your Knowledge Core

Identify the knowledge that professionals in your field must possess instantly to reason effectively under uncertainty.

Create a short list of foundational concepts, models, or facts that should be internalized rather than merely accessible.

Treat this list as the **core knowledge base** your thinking depends upon.

The goal is not to memorize everything.

It is to ensure that the knowledge required for reasoning remains inside the mind rather than entirely outside it.

CHAPTER 10

ANALYSIS VERSUS JUDGMENT

WHERE HUMAN REASONING REMAINS ESSENTIAL

Late one evening, a wildfire operations team gathers around a map table.

Satellite data, wind forecasts, and terrain models cover the screen in front of them. AI-assisted systems have already analyzed thousands of variables: temperature gradients, humidity levels, vegetation density, historical burn patterns.

The models generate predictions about how the fire will spread overnight.

Most of the projections agree.

The fire will likely continue advancing northeast.

Containment lines have been placed accordingly.

But one of the incident commanders studies the terrain map quietly.

Something about the pattern does not sit right.

The data shows stable winds.

But the commander knows the canyon system in this region can create sudden directional shifts when nighttime cooling begins.

The models do not show that change.

Because it has not occurred often enough in the historical data to dominate the statistical pattern.

The commander orders crews to reposition equipment along a ridge the models considered low risk.

Some members of the team hesitate.

The system's projections suggest the adjustment is unnecessary.

But the order stands.

At three o'clock in the morning, the wind shifts.

The fire moves directly toward the ridge the commander anticipated.

Because crews were already positioned there, containment holds.

The decision prevented a major expansion of the fire.

Nothing about the decision contradicted the analysis.

The commander simply recognized where the analysis stopped.

And where judgment had to begin.

The Governing Distinction

The previous chapters examined how AI systems reshape professional cognition.

They showed how fluent outputs can distort trust, how retrieval can replace generation, how authority can transfer from thinker to system, and how organizations can converge around shared analytical frameworks.

But none of those chapters addressed the deeper question now facing professionals:

What role does human thinking still play when machines can perform analysis at extraordinary scale?

The answer begins with a fundamental distinction.

Analysis versus judgment.

Analysis identifies patterns within known data.

Judgment decides what to do when those patterns are incomplete.

Analysis organizes evidence.

Judgment acts when evidence runs out.

Analysis is computational.

Judgment is contextual.

AI systems excel at analysis.

Human reasoning remains essential for judgment.

Where Analysis Ends

In many professional settings, analysis can now be performed faster and more comprehensively than any individual could manage alone.

But the moment where analysis ends appears in familiar places.

A physician reviewing diagnostic probabilities must still decide whether a patient's symptoms represent a common condition presenting unusually or a rare condition presenting early.

An attorney analyzing case law through an AI system may receive a well-structured summary of precedent but must still judge how a particular judge is likely to interpret those precedents in the specific courtroom context.

A financial analyst may receive sophisticated model forecasts yet still decide whether the assumptions behind those forecasts remain valid in a rapidly shifting market.

In each case analysis organizes available information.

Judgment decides how that information applies to a situation that may not perfectly resemble the past.

The Nature of Judgment

Judgment is often misunderstood as intuition or instinct.

In reality, strong judgment is rarely spontaneous.

It emerges from compressed knowledge, experience, and situational awareness interacting in real time.

Judgment asks questions analysis cannot answer.

What assumptions might be wrong?

What signals fall outside the data?

What consequences matter most if the model fails?

What decision preserves the most flexibility if conditions change?

These questions rarely produce precise answers.

They shape the direction of action.

Judgment does not replace analysis.

It determines how analysis should be used.

The Illusion of Complete Analysis

One of the subtle effects of powerful analytical systems is the sense that decisions should wait until analysis becomes complete.

But in many situations, analysis can expand indefinitely.

More data can always be gathered.

More scenarios can be modeled.

More forecasts can be generated.

Judgment provides the structure that allows action within that complexity.

It asks which signals matter.

It prioritizes competing risks.

It decides when analysis is sufficient and when additional information will not meaningfully change the decision.

In this sense, the purpose of human judgment is not to compete with AI analysis.

It is to govern it.

Decision-Making Without Precedent

The most difficult decisions occur when historical precedent offers little guidance.

These are the situations where AI analysis becomes least reliable.

Pandemic responses early in a novel outbreak.

Financial decisions during unprecedented market conditions.

Technological choices involving innovations that have never been deployed at scale.

In these environments the available data may be incomplete or misleading.

Models may offer projections based on weak assumptions.

Waiting for perfect analysis may itself create risk.

Judgment becomes the ability to act responsibly without certainty.

It balances evidence, experience, and consequences.

And it accepts that some decisions must be made before analysis is complete.

A Return to the Physician

The physician in Chapter 9 faced a similar boundary.

Diagnostic systems provided valuable analysis.

They summarized research and suggested possible explanations for the patient's symptoms.

But if the patient's condition deteriorated rapidly, the physician could not pause indefinitely to retrieve additional information.

The decision had to be made using knowledge already internalized and situational signals unfolding in real time.

The system contributed analysis.

The physician supplied judgment.

That relationship between analysis and judgment appears across professional life.

Analysis expands what we can see.

Judgment determines what we do.

From Diagnosis to Design

The previous nine chapters diagnosed how AI systems can quietly reshape professional thinking.

They showed the risks of fluency distortion, authority transfer, reactive attention, institutional convergence, and knowledge fragmentation.

Chapter 10 marks a transition.

The goal is no longer simply to understand what goes wrong.

The goal is to design thinking systems that combine the strengths of analysis and judgment.

AI systems expand analytical capacity.

Human cognition provides contextual understanding, ethical responsibility, and the ability to act under uncertainty.

When these roles are clearly understood, the relationship between human and machine becomes collaborative rather than competitive.

This collaboration forms the foundation of the framework introduced in the next chapter.

Exercises

Exercise 1: Identifying the Boundary

Think about a recent decision you made that involved significant uncertainty.

List the parts of the decision that could be addressed through analysis.

Then identify the point where analysis stopped and judgment began.

Understanding this boundary helps clarify the role each form of thinking plays.

Exercise 2: Stress Testing Analysis

Take an analytical conclusion produced by an AI system or data model.

Ask three questions:

1. What assumptions does this analysis rely on?
2. What conditions would make those assumptions fail?
3. What would the consequences be if the model were wrong?

This exercise trains judgment to examine analysis rather than accept it automatically.

Exercise 3: Applying the Judgment Questions

Take a recent AI-assisted decision you made.

Now apply the four judgment questions from this chapter:

- What assumptions might be wrong?
- What signals fall outside the data?
- What consequences matter most if the model fails?
- What decision preserves the most flexibility if conditions change?

Reflect on whether the decision would have changed if these questions had been asked earlier.

Judgment improves not by replacing analysis, but by learning how to interrogate it.

CHAPTER 11

THE AMPLIFIED MIND

SUBSTITUTION VERSUS AMPLIFICATION

A litigation team gathers late in the evening to prepare for a major hearing the following morning.

The lead attorney has been reviewing thousands of pages of depositions, financial records, and internal communications for weeks. The case is complex. Several regulatory questions overlap with contract law, and the opposing counsel is known for introducing unexpected arguments.

The attorney opens an AI research system.

Instead of asking the system to generate a summary of the case, she begins differently.

First she outlines her own theory of the argument.

She sketches the structure on a notepad.

Two central claims.

Three supporting precedents.

One weakness the opposing side is likely to exploit.

Only after building that structure does she begin interacting with the system.

She asks it to locate cases that challenge her argument.

The system returns several possibilities.

Two are irrelevant.

One is concerning.

She studies it carefully.

Then she asks the system to test her reasoning.

"Assume the opposing counsel frames the issue this way. What argument would they make?"

The system generates a plausible counterargument.

She adjusts her reasoning.

Then she asks a different question.

"Where does my logic depend on assumptions that may not hold?"

The system identifies two areas of vulnerability.

She revises again.

Over the next hour the interaction continues.

The system expands the search space.

The attorney expands the reasoning.

Neither replaces the other.

By midnight the argument has become stronger than either the attorney or the system could have produced independently.

The system supplied analysis.

The attorney supplied structure, judgment, and direction.

Together they produced something neither could generate alone.

The Difference You Just Saw

For ten chapters this book has shown what happens when AI systems quietly reshape human cognition.

We have seen professionals accept fluent answers too quickly.

We have seen retrieval replace generation.

We have seen authority transfer from thinker to system.

We have seen organizations converge around shared analytical frameworks.

We have seen knowledge migrate outward into external systems.

In each case the pattern was similar.

Human cognition slowly moved out of the center of the thinking process.

The scene you just read works differently.

The system did not replace the attorney's reasoning.

It expanded it.

The attorney generated the structure.

The system expanded the search.

The attorney evaluated the results.

The system tested the assumptions.

The reasoning remained human.

The analysis became superhuman.

This is the distinction the entire book has been building toward.

The Governing Distinction

Substitution versus amplification.

Substitution occurs when AI systems replace the cognitive work that builds expertise and judgment.

Amplification occurs when AI systems expand what human cognition can do without replacing the thinking that produces understanding.

Substitution makes humans more efficient.

Amplification makes humans more capable.

Substitution removes cognitive friction.

Amplification directs cognitive friction toward deeper reasoning.

Substitution answers questions.

Amplification helps thinkers ask better ones.

Both approaches use the same technology.

Only one strengthens human intelligence over time.

Why Substitution Is Easy

Substitution is attractive because it appears efficient.

If a system can produce a fluent answer instantly, generating the reasoning yourself can feel unnecessary.

Each of these moments saves time.

But they gradually move the center of cognition outward.

The professional becomes a reviewer of generated reasoning rather than the originator of it.

Substitution often looks like productivity.

But over time it erodes the very capabilities professionals rely on when conditions change.

Why Amplification Requires Discipline

Amplification does not occur automatically.

It requires maintaining the cognitive practices that build expertise while using AI systems to expand analytical capacity.

Generation before retrieval.

Internal knowledge before external access.

Judgment governing analysis rather than following it.

These practices feel slower in the moment.

But they preserve the structures of thinking that allow professionals to use AI systems intelligently rather than passively.

Amplification therefore depends on discipline.

Not technological discipline.

Cognitive discipline.

The discipline to remain the originator of reasoning even when systems can generate reasoning instantly.

The Structure of an Amplified Mind

When amplification is working, the relationship between human cognition and AI systems follows a recognizable pattern.

The human thinker generates the initial structure of the problem.

The AI system expands the analytical search.

The human evaluates the results and reframes the question.

The system tests assumptions and identifies additional possibilities.

The human integrates the new information into a revised model of the situation.

This cycle repeats.

Each pass expands the space of possible reasoning while preserving human judgment as the organizing force.

The system supplies scale.

The human supplies direction.

Over time the result is not simply faster thinking.

It is **larger thinking**.

The Role of Cognitive Friction

Many of the diagnostic chapters described how AI systems remove friction from thinking processes.

Answers appear instantly.

Research is summarized automatically.

Writing becomes effortless.

In an amplified system, friction is not eliminated.

It is repositioned.

Instead of spending effort searching for information, professionals spend effort refining questions, evaluating assumptions, and integrating insights.

The work shifts from gathering data to shaping understanding.

That shift is the difference between substitution and amplification.

Substitution removes effort from thinking.

Amplification moves effort toward the parts of thinking that matter most.

The Model Emerging

The patterns described across the previous chapters now begin to form a coherent structure.

Generation builds knowledge.

Knowledge compression enables judgment.

Judgment governs analysis.

Analysis expands the space of possible reasoning.

Each element depends on the others.

Remove generation and knowledge weakens.

Remove knowledge and judgment falters.

Remove judgment and analysis loses direction.

Remove analysis and human reasoning remains limited by scale.

The Amplified Mind Model emerges from maintaining the full cycle rather than allowing any part of it to disappear.

AI systems expand cognition.

Human discipline preserves it.

The Opportunity Ahead

The arrival of powerful analytical systems does not inevitably weaken human thinking.

It creates a choice.

Professionals and organizations can allow substitution to quietly displace the cognitive work that builds expertise.

Or they can design workflows that amplify human reasoning.

The difference lies not in the technology itself.

It lies in how the technology is used.

Amplification means remaining the originator of thought even while machines expand what that thought can reach.

The attorney at the beginning of this chapter did not become less intelligent by using the system.

Her intelligence became larger.

Because the system expanded her reasoning without replacing it.

That is what amplified intelligence feels like in practice.

Exercises

Exercise 1: Generation First

The next time you approach a complex problem, write down your reasoning before consulting any AI system.

Outline your understanding of the issue.

List possible explanations or strategies.

Only afterward ask the system to expand, challenge, or test your thinking.

This preserves generation as the starting point of amplification.

Exercise 2: Assumption Testing

After receiving an AI-generated analysis, deliberately challenge it.

Ask the system to identify weaknesses in its own reasoning.

Ask for counterarguments.

Ask what evidence would invalidate the conclusion.

Amplification improves when analysis is treated as material for reasoning rather than as an answer.

Exercise 3: Designing an Amplified Workflow

Choose one recurring task in your professional life.

Redesign the workflow so that:

1. Human reasoning generates the initial structure.

2. AI systems expand the analytical possibilities.

3. Human judgment evaluates and integrates the results.

The goal is to ensure the system expands thinking rather than replacing it.

CHAPTER 12

DESIGNING AMPLIFIED WORKFLOWS

ACCIDENTAL VERSUS INTENTIONAL DESIGN

The Decision the Organization Didn't Realize It Had Made

Two consulting teams in the same firm adopted the same AI system within the same month.

Both teams were staffed with experienced analysts. Both had access to the same data. Both were tasked with producing weekly strategic briefs for clients.

Six months later their work looked very different.

The first team used the system the way most organizations do. When a new project arrived, the analysts opened the AI interface and asked it to draft an outline of the problem. They asked it to summarize background material. They asked it to generate possible explanations and recommendations.

The system produced fluent responses quickly. Drafts appeared in minutes. Research summaries arrived instantly.

The analysts edited what the system produced, reorganized sections, and added details from their own experience.

The process was efficient. Work moved faster than before.

But something subtle began to change.

The analysts gradually stopped generating their own initial structures for problems. They stopped sketching rough frameworks before opening the system. They stopped asking what explanation made sense before asking the AI for possibilities.

The system provided those things immediately.

At first the change felt like progress.

But when unusual client problems appeared, the analysts struggled to recognize weak arguments in the AI drafts. They struggled to see which sources mattered more than others. They struggled to decide which explanation actually fit the situation.

They had become excellent editors of AI output.

But the underlying work of generating explanations had quietly moved elsewhere.

The second team adopted the same system.

But their workflow began differently.

Before anyone opened the AI interface, the analyst responsible for the brief wrote three things on a notepad.

The central claim they believed might explain the problem.
Two alternative explanations that might challenge that claim.
And one piece of evidence that could potentially weaken the argument.

Only after writing those things did they consult the AI system.

They did not ask it to produce an answer.

They asked it to test the one they had already begun.

Where might this argument break down?
What evidence contradicts this explanation?
What sources challenge these assumptions?

The system responded the way AI systems do, rapidly and fluently.

But the analyst was not asking the system to think for them.

They were asking it to pressure test the thinking they had already begun.

Over time the difference between the two teams became visible.

The first team produced faster drafts.
The second team produced better reasoning.

Both were using the same tool.

Only one redesigned the workflow.

The Difference Between Using AI and Designing With It

What you just saw was not a difference in technology.

It was a difference in design.

One organization adopted a powerful tool and inserted it into existing workflows without changing how thinking happened inside those workflows.

The other redesigned the workflow itself.

The distinction is simple.

Accidental systems emerge when tools are added without redesigning the process around them.
Intentional systems emerge when workflows are engineered so human and machine strengths reinforce each other.

Most organizations today operate accidental systems.

They adopt powerful tools and assume improved capability will follow automatically.

But tools alone do not determine outcomes.

Workflows do.

Why Design Matters

Substitution does not usually appear because people intend it.

It appears because workflows make it easy.

When a system can produce an outline instantly, generating one yourself feels unnecessary.

When a system can summarize sources immediately, reading primary material feels inefficient.

When a system can suggest explanations quickly, asking your own questions feels redundant.

Over time the structure of thinking shifts.

Generation becomes retrieval.
Exploration becomes editing.
Judgment becomes selection among options already produced.

Preventing that shift requires redesigning the workflow itself.

Principle 1

Generation Comes Before Retrieval

In every amplified workflow, the human thinker generates the first structure of the problem before consulting the system.

This structure does not need to be polished.

It can be a rough outline.
A hypothesis.
A sketch of an argument.

What matters is that the initial frame originates from human reasoning.

Once that structure exists, the AI system becomes powerful in a different way.

It can challenge the structure.
Expand it.
Test it against information far beyond the individual's reach.

But the direction of thinking remains human-generated.

Without this step, the system becomes the source of the structure itself.

And when that happens, the human role shifts from creator to editor.

Generation is the moment where the thinker decides what question is being asked.

If that moment disappears, the rest of the workflow changes with it.

Principle 2

AI Interaction Is Structured as Critique

In substitution workflows, AI is asked to produce answers.

In amplified workflows, AI is asked to test them.

Instead of asking:

"Write an explanation of this problem."

The analyst asks:

"Where might this explanation fail?"

Instead of asking:

"Generate an outline for this report."

They ask:

"What assumptions in this outline are weakest?"

The system still produces ideas, evidence, and alternatives.

But the human thinker remains responsible for the original reasoning.

The system becomes a powerful critic rather than a substitute author.

This is where AI systems are uniquely valuable.

They can expose blind spots rapidly.
Surface counterexamples instantly.
Expand the space of possibilities beyond what any single thinker could explore alone.

Amplification happens when critique strengthens human reasoning.

Principle 3

Judgment Checkpoints Are Built Into the Workflow

Amplified workflows include specific moments where human evaluation is required before work continues.

These checkpoints prevent the quiet drift toward substitution.

After the system produces additional evidence, the analyst must decide:

Does this information actually strengthen the argument?

After the system generates alternative explanations, the analyst must determine:

Which explanation best fits the evidence?

After the system suggests revisions, the analyst must evaluate:

Which changes improve the reasoning and which simply make the text more fluent?

Without these moments, AI output flows directly into final work products.

With them, human judgment remains central to the process.

The checkpoint is where thinking happens.

It is the moment where the thinker integrates new information into an evolving understanding.

That integration cannot be automated.

What an Amplified Workflow Looks Like

Consider a common professional task: preparing a strategic memo.

A substitution workflow often looks like this:

1. Ask AI to summarize background information.

2. Ask AI to propose explanations.
3. Ask AI to draft the memo.
4. Edit the result.

The system does most of the reasoning.

An amplified workflow looks different.

1. Generate a rough explanation of the situation.
2. Ask AI to identify weaknesses, counterexamples, and missing evidence.
3. Evaluate the critique and refine the argument.
4. Ask AI for additional evidence or alternative interpretations.
5. Decide which explanation best fits the evidence.
6. Draft the memo using the refined reasoning.

The system expands the information available and challenges assumptions.

But the core reasoning remains human.

The AI system becomes a force multiplier rather than a replacement.

Designing This on Monday Morning

Amplified workflows do not require new tools.

They require small design decisions.

Write the first hypothesis yourself before opening the system.

Ask AI for critique rather than answers.

Insert moments where you must decide what the output actually means.

Those three changes move cognitive effort toward the parts of thinking that matter most.

Exercises

Exercise 1: Generation and Retrieval

Choose a recurring task in your work that currently begins with an AI prompt.

Before consulting the system, write down:

- Your initial hypothesis about the problem.
- Two alternative explanations that could challenge it.
- One piece of evidence that could weaken your argument.

Only after generating these should you consult the AI system.

Notice how the system's responses change when they are testing your thinking rather than replacing it.

Exercise 2: Turning AI Into a Critic

Take a prompt you regularly use with AI.

Rewrite it so the system critiques your reasoning rather than generating it.

Example:

Substitution prompt:
"Explain the causes of declining customer retention."

Amplified prompt:
"Here is my explanation for declining customer retention. Where might this reasoning break down?"

Repeat this with several prompts you commonly use.

Observe how the system's role shifts when it is asked to challenge rather than create.

Exercise 3: Designing a Judgment Checkpoint

Select a workflow you perform regularly, writing a report, preparing a presentation, analyzing data.

Identify one moment where AI output currently flows directly into the next step.

Insert a deliberate checkpoint.

At that moment, ask yourself:

- Does this output actually strengthen my reasoning?
- What assumptions does it rely on?
- What might it be missing?

Design the workflow so that this evaluation must occur before the task continues.

This checkpoint is where amplification happens.

The System You Are Actually Designing

Every organization adopting AI tools today is designing a thinking system.

Most do not realize it.

They believe they are simply adding productivity tools to existing processes.

But workflows shape cognition.

They determine where effort occurs, where questions originate, and where judgment is required.

An accidental workflow gradually transfers these functions to the system itself.

An intentional workflow strengthens them.

The technology is the same in both cases.

The design is what changes the outcome.

Human intelligence does not shrink in the presence of powerful tools.

It expands.

CHAPTER 13

DESIGNING ORGANIZATIONAL INTELLIGENCE

SYNCHRONIZED THINKING VERSUS COLLECTIVE INTELLIGENCE

The Meeting That Could Have Gone Either Way

The executive team of a large financial services firm gathered for their quarterly strategy session.

Over the previous year the firm had invested heavily in AI systems. Analysts across the organization now used the same tools to summarize market data, generate research briefs, and evaluate strategic scenarios. Productivity had improved quickly. Reports arrived faster. Forecasting models could be updated in minutes.

But something had begun to worry the firm's chief strategy officer.

Across divisions, analysts were producing remarkably similar conclusions.

Different teams working independently on the same market questions were arriving at nearly identical frameworks. The language of their reports sounded the same. The categories used to analyze problems were almost interchangeable.

At first this consistency looked like alignment.

But the strategy officer suspected something else was happening.

The organization was not thinking together.

It was thinking the same.

That observation shaped the agenda for the meeting.

How should the firm structure AI-assisted decision making going forward?

One option was the path many organizations were already taking.

Standardize the tools.
Standardize the prompts.
Standardize the analytical frameworks produced by the system.

Ensure that every team used the same models to evaluate opportunities and risks.

The advantages were obvious.

Decisions would move faster.
Reports would be consistent.
Strategic alignment would be easy to maintain.

But the strategy officer posed a different question.

What happens when everyone in the organization begins with the same analytical frame?

The room was quiet for a moment.

Because the answer was clear.

Disagreement would shrink.

Not because the problems were simple.

But because everyone was starting from the same assumptions.

The board chair leaned forward.

"If the system gives everyone the same framework," she said, "we may become more efficient."

She paused.

"But we may also become less intelligent."

The Difference Between Agreement and Intelligence

Chapter 8 described what happens when institutions unknowingly converge around shared analytical frameworks.

Different people ask similar questions.
Use the same conceptual categories.
Reach conclusions that differ in detail but not in structure.

This pattern produces what looks like diversity.

But underneath the surface the reasoning is synchronized.

The same analytical lens is being applied everywhere.

This is **synchronized thinking**.

It feels like alignment.

But alignment achieved through convergence reduces the number of perspectives available to an organization.

When uncertainty increases, that reduction becomes dangerous.

The alternative is harder to maintain.

It requires institutions to preserve multiple ways of thinking about the same problem at the same time.

This is **collective intelligence**.

In a collectively intelligent organization:

Multiple ways of framing a problem exist simultaneously.

Different teams approach the same question through different analytical lenses.

Judgment remains distributed across people rather than concentrated in system outputs.

The goal is not agreement.

The goal is a richer understanding of the problem.

Synchronization is efficient.

Collective intelligence is not.

But intelligence has never been the product of efficiency alone.

The Design Problem Institutions Face

AI systems produce coherent analytical frameworks quickly.

When many people in an organization consult the same systems, those frameworks naturally begin to converge.

Without deliberate design, institutional thinking synchronizes.

Preventing that outcome requires designing the organization's decision processes differently.

The three principles that follow translate the Amplified Mind Model from individual practice into institutional design.

Principle 1

Distributed Generation

In organizations that preserve collective intelligence, the initial framing of important problems occurs across multiple independent perspectives before AI systems are consulted.

Instead of beginning with a single AI-generated framework, several teams generate their own initial explanations.

Each group asks:

What do we believe is happening here?

What evidence supports that view?

What alternative explanations might exist?

Only after those perspectives are articulated do teams consult AI systems to test and expand them.

This design preserves the generative step across the organization.

Different thinkers begin from different assumptions.

Different interpretations of the evidence emerge.

The system then becomes a tool for expanding and challenging those perspectives, rather than replacing them.

Distributed generation ensures that the organization begins with multiple ways of seeing the problem.

Without it, everyone begins with the same one.

Principle 2

Structured Dissent

In synchronized organizations, disagreement fades gradually.

Not because people stop thinking critically, but because everyone begins from the same analytical frame.

Collective intelligence requires deliberate mechanisms for challenging that frame.

This is **structured dissent**.

After an AI-assisted analysis is produced, specific individuals or teams are tasked with challenging it.

Their role is not to improve the analysis.

Their role is to question it.

They ask:

What assumptions does this framework rely on?

What alternative interpretations of the evidence exist?

What would this analysis look like if the underlying model were wrong?

By institutionalizing dissent, organizations ensure that machine-generated frameworks do not become default truths.

They remain hypotheses.

And hypotheses can be challenged.

Principle 3

Accountability for Judgment

AI systems can generate options.

They cannot assume responsibility for decisions.

In synchronized organizations, responsibility gradually diffuses.

When everyone relies on the same system output, it becomes unclear who actually owns the final judgment.

The analysis becomes "what the model says."

Collective intelligence requires the opposite.

Specific humans remain accountable for specific decisions.

The system can inform the decision.

It can expand the available evidence.

But a person must ultimately say:

This is the conclusion we accept.

And they must be responsible for that judgment.

Accountability preserves the role of human reasoning in institutional decision making.

Without it, authority quietly shifts to the system itself.

What Institutional Amplification Looks Like

Imagine a major strategic decision facing an organization.

In a synchronized institution, the process often looks like this:

The leadership team consults an AI system.
The system produces a comprehensive analysis.
The team discusses the results and selects the most plausible recommendation.

The analysis feels objective because it originates from the same system everyone trusts.

But the range of perspectives is narrow.

Everyone began with the same analytical structure.

In a collectively intelligent institution, the process looks different.

Multiple teams independently frame the problem and generate initial explanations.

Each perspective is documented before AI systems are consulted.

The system is then used to test those competing interpretations, surfacing additional evidence and alternative scenarios.

A separate group is tasked with challenging the dominant framework.

Only after these steps does leadership evaluate the competing analyses and make a decision.

The organization does not produce a single answer immediately.

It produces a deeper understanding of the problem.

The system expands the analysis.

Human judgment integrates it.

Scaling Amplification

Chapter 12 showed how an individual professional can redesign their workflow to amplify thinking.

Chapter 13 shows the same logic applied at institutional scale.

Generation occurs across multiple minds rather than a single one.

Critique becomes structured dissent.

Judgment remains anchored in accountable human decision makers.

These are not technological choices.

They are design choices.

Organizations that adopt AI without redesigning their thinking systems will gradually synchronize around machine-generated frameworks.

Organizations that design deliberately can achieve something different.

They can combine the analytical power of machines with the diversity of human reasoning.

Exercises

Exercise 1: Mapping Generation in Your Organization

Choose a recent strategic question your organization evaluated.

Trace how the problem was initially framed.

Who defined the problem?
How many independent perspectives existed before AI systems or analytical tools were consulted?
Did multiple teams generate competing interpretations, or did a single framework dominate the discussion?

Map this process on paper.

Then ask: how could distributed generation be introduced so multiple perspectives emerge before systems are consulted?

Exercise 2: Practicing Structured Dissent

Select a recent analysis produced with AI assistance.

Assign one person or team the explicit role of challenger.

Their responsibility is to question the analysis, not refine it.

Ask them to identify:

- The assumptions embedded in the framework
- Alternative interpretations of the same evidence
- Scenarios where the dominant explanation would fail

Observe how the analysis changes when dissent is intentionally structured rather than informally expressed.

Exercise 3: Designing Accountability for Judgment

Identify a recent decision where the reasoning behind the outcome was described as "what the model showed" or "what the analysis suggested."

Ask three questions:

Who was responsible for interpreting the analysis?
Who owned the final judgment?
Was that responsibility clearly defined before the decision process began?

For the next similar decision, explicitly assign a decision owner whose responsibility is to interpret the analysis and make the final judgment.

The system informs the decision.

The human owns it.

The Intelligence of an Institution

Institutions do not think the way individuals do.

Their intelligence emerges from how many minds interact with one another.

AI systems will shape those interactions.

But they do not determine them.

That is a design decision.

An organization can allow its thinking to synchronize around a single analytical lens.

Or it can deliberately cultivate a system where multiple perspectives remain visible and judgment remains human.

One path produces agreement.

The other produces intelligence.

CHAPTER 14

THE DISCIPLINE OF AMPLIFIED THINKING

THE CHOICE THAT APPEARS INVISIBLE

A young analyst sits down to begin a report.

The question in front of her is difficult. The evidence is incomplete. The explanation is not obvious.

She opens her laptop.

The cursor blinks in an empty prompt field.

At that moment she has two options.

She can ask the system to explain the problem.

Or she can begin thinking.

The difference between those choices will not be visible in the final document. The report will read smoothly either way. The argument may even look similar.

But the two paths lead to very different outcomes.

One produces efficiency.

The other produces intelligence.

The difference between them is the central argument of this book.

The Question the Book Set Out to Answer

The debate surrounding artificial intelligence has often been framed in simple terms.

Is AI good for human thinking or bad for it?

Will it strengthen cognition or weaken it?

Those questions assume the answer lies in the technology itself.

The argument of this book has been different.

Technology does not determine cognitive outcomes.

Design does.

The same system can produce substitution or amplification depending on how it is integrated into human reasoning.

Across the chapters of this book we have examined the mechanisms that shape that outcome.

How ideas are generated.

How knowledge is compressed.

How judgment operates.

How analysis scales.

How institutions converge.

How workflows either remove or redirect cognitive effort.

Each distinction identified a pressure point where thinking can shift.

Taken together they form something larger than a list of observations.

They form a theory of thinking in an AI-assisted environment.

The Theory in One Sentence

Human intelligence in the presence of AI is determined not by the capabilities of the system, but by the discipline of the thinker.

When generation disappears, knowledge weakens.

When knowledge weakens, judgment falters.

When judgment falters, analysis becomes mechanical.

When analysis becomes mechanical, reasoning follows the path the system provides.

But when the generative step is preserved, something different happens.

The system expands the reach of human thought without replacing its source.

This is the distinction between substitution and amplification.

It is the governing choice beneath every interaction between a thinker and a machine.

Why Discipline Matters

Machines remove friction.

That is one of their greatest strengths.

They accelerate retrieval, summarize information, generate text, and explore possibilities at a speed no human mind can match.

But friction has always played an important role in thinking.

The effort required to generate an explanation forces a thinker to organize knowledge.

The effort required to read primary sources builds understanding.

The effort required to question assumptions sharpens judgment.

When that friction disappears completely, these capabilities gradually weaken.

Amplification does not eliminate friction.

It moves it.

Instead of spending effort gathering information, the thinker spends effort shaping understanding.

Instead of searching endlessly for sources, the thinker evaluates their meaning.

Instead of generating endless drafts, the thinker asks better questions.

The discipline lies in choosing where effort remains.

The Practices That Sustain Amplification

The previous chapters translated this discipline into practical design.

At the individual level, amplified thinking requires three habits.

Generation before retrieval.
Critique rather than delegation.
Judgment checkpoints where human evaluation remains central.

At the institutional level, the same logic scales upward.

Distributed generation preserves multiple ways of framing problems.

Structured dissent protects organizations from synchronized thinking.

Accountability for judgment ensures that responsibility remains human.

These practices are not technological innovations.

They are cognitive ones.

They determine whether AI becomes a substitute for thinking or an amplifier of it.

The Future of Thinking

Every generation encounters tools that reshape how knowledge is created and shared.

The printing press expanded access to ideas.

The internet expanded access to information.

Artificial intelligence expands access to reasoning itself.

For the first time, machines can produce explanations, arguments, and analyses that resemble human thought.

This capability creates a temptation unlike any previous technological shift.

The temptation is not merely to automate labor.

It is to automate thinking.

Whether that temptation is resisted will determine the intellectual character of the coming decades.

If thinking is delegated completely, human reasoning will narrow.

People will become excellent editors of machine-generated work.

But if thinking is amplified, the opposite can occur.

Human intelligence can expand beyond the limits of individual cognition.

Machines can extend the reach of reasoning while humans retain responsibility for direction, interpretation, and judgment.

The Discipline of the Amplified Mind

An amplified mind is not one that relies on machines.

It is one that knows how to work with them.

It begins with a question rather than a prompt.

It generates an explanation before asking for one.

It treats machine output as a hypothesis rather than an answer.

It evaluates, challenges, and refines what the system produces.

The machine becomes a collaborator in exploration.

But the thinker remains responsible for understanding.

This discipline is subtle.

No rule forces it.

No interface requires it.

Every interaction with an intelligent system contains the same quiet choice.

To think first.

Or to ask the machine to think instead.

The Intelligence That Remains Human

The most important decisions people make rarely arrive with complete information.

They involve uncertainty, judgment, and competing interpretations of the same evidence.

Machines can help explore those interpretations.

They cannot decide which ones matter most.

That responsibility remains human.

It belongs to the person asking the question, the professional making the judgment, the institution choosing a course of action.

Technology may extend the reach of human reasoning.

But meaning, responsibility, and understanding remain human tasks.

The system informs the thinking.

The human owns the conclusion.

The Quiet Choice

The future of human intelligence will not be determined by machines alone.

It will be determined by millions of small decisions made every day by people interacting with them.

The analyst deciding whether to generate an explanation before asking for one.

The researcher choosing whether to read a source before summarizing it.

The organization deciding whether to preserve dissent or standardize its thinking.

These choices rarely feel dramatic.

But they accumulate.

Over time they determine whether human reasoning narrows or expands.

The tools will continue to improve.

They will grow faster, more capable, and more persuasive.

But the central question will remain unchanged.

Will humans use these systems to avoid thinking?

Or to extend it?

The answer will not come from the machines.

It will come from the discipline of the minds that use them.

NOTES

Chapter 1: Fluency Distortion

1. Norbert Schwarz, "Metacognitive Experiences in Consumer Judgment and Decision Making," Journal of Consumer Psychology 14, no. 4 (2004): 332–348.

2. Daniel Kahneman, Thinking, Fast and Slow (New York: Farrar, Straus and Giroux, 2011).

3. Raja Parasuraman and Victor Riley, "Humans and Automation: Use, Misuse, Disuse, Abuse," Human Factors 39, no. 2 (1997): 230–253.

Chapter 2: Cognitive Offloading

4. Betsy Sparrow, Jenny Liu, and Daniel M. Wegner, "Google Effects on Memory: Cognitive Consequences of Having Information at Our Fingertips," Science 333, no. 6043 (2011): 776–778.

5. Daniel M. Wegner, "Transactive Memory: A Contemporary Analysis of the Group Mind," in Theories of Group Behavior, ed. Brian Mullen and George R. Goethals (New York: Springer-Verlag, 1987), 185–208.

6. John Sweller, "Cognitive Load During Problem Solving: Effects on Learning," Cognitive Science 12, no. 2 (1988): 257–285.

7. Anders Ericsson and Robert Pool, Peak: Secrets from the New Science of Expertise (Boston: Houghton Mifflin Harcourt, 2016).

Chapter 3: Authority Transfer

8. Raja Parasuraman and Victor Riley, "Humans and Automation: Use, Misuse, Disuse, Abuse," Human Factors 39, no. 2 (1997): 230–253.

9. Berkeley J. Dietvorst, Joseph P. Simmons, and Cade Massey, "Algorithm Aversion: People Erroneously Avoid Algorithms After Seeing Them Err," Journal of Experimental Psychology: General 144, no. 1 (2015): 114–126.

10. Jennifer M. Logg, Julia A. Minson, and Don A. Moore, "Algorithm Appreciation: People Prefer Algorithmic to Human Judgment," Organizational Behavior and Human Decision Processes 151 (2019): 90–103.

11. Daniel Kahneman, Thinking, Fast and Slow (New York: Farrar, Straus and Giroux, 2011).

Chapter 4: Reactive Minds

12. Herbert A. Simon, "Designing Organizations for an Information-Rich World," in Computers, Communications, and the Public Interest, ed. Martin Greenberger (Baltimore: Johns Hopkins Press, 1971).

13. Sophie Leroy, "Why Is It So Hard to Do My Work? The Challenge of Attention Residue When Switching Between Work Tasks," Organizational Behavior and Human Decision Processes 109, no. 2 (2009): 168-181.

14. Gloria Mark, Victor M. Gonzalez, and Justin Harris, "No Task Left Behind? Examining the Nature of Fragmented Work," Proceedings of the SIGCHI Conference on Human Factors in Computing Systems (2005).

15. Cal Newport, Deep Work: Rules for Focused Success in a Distracted World (New York: Grand Central Publishing, 2016).

Chapter 5: The Strain

16. John Sweller, "Cognitive Load During Problem Solving: Effects on Learning," Cognitive Science 12, no. 2 (1988): 257–285.

17. Anders Ericsson and Robert Pool, Peak: Secrets from the New Science of Expertise (Boston: Houghton Mifflin Harcourt, 2016).

18. Elizabeth L. Bjork and Robert A. Bjork, "Making Things Hard on Yourself, But in a Good Way: Creating Desirable Difficulties to Enhance Learning," in Psychology and the Real World, ed. Morton Ann Gernsbacher et al. (New York: Worth Publishers, 2011).

19. Gary Klein, Sources of Power: How People Make Decisions (Cambridge, MA: MIT Press, 1998).

Chapter 6: Authorship

20. John M. Darley and Bibb Latané, "Bystander Intervention in Emergencies: Diffusion of Responsibility," Journal of Personality and Social Psychology 8, no. 4 (1968): 377–383.

21. Leonid Rozenblit and Frank C. Keil, "The Misunderstood Limits of Folk Science: An Illusion of Explanatory Depth," Cognitive Science 26, no. 5 (2002): 521–562.

22. Daniel M. Wegner, The Illusion of Conscious Will (Cambridge, MA: MIT Press, 2002).

23. The Feynman Technique refers to a learning method derived from Richard Feynman's philosophy of understanding through explanation, widely discussed in educational writing. For Feynman's approach to learning, see James Gleick, Genius: The Life and Science of Richard Feynman (New York: Pantheon Books, 1992).

Chapter 7: The Signal Problem

24. Herbert A. Simon, "Designing Organizations for an Information-Rich World," in Computers, Communications, and the Public Interest, ed. Martin Greenberger (Baltimore: Johns Hopkins Press, 1971).

25. Cal Newport, Deep Work: Rules for Focused Success in a Distracted World (New York: Grand Central Publishing, 2016).

26. Gloria Mark, Attention Span: A Groundbreaking Way to Restore Balance, Happiness, and Productivity (New York: Hanover Square Press, 2023).

27. Mihaly Csikszentmihalyi, Flow: The Psychology of Optimal Experience (New York: Harper & Row, 1990).

Chapter 8: Institutional Convergence

28. Cass R. Sunstein, Infotopia: How Many Minds Produce Knowledge (Oxford: Oxford University Press, 2006). On informational cascades, see also Sushil Bikhchandani, David Hirshleifer, and Ivo Welch, "A Theory of Fads, Fashion, Custom, and Cultural Change as Informational Cascades," Journal of Political Economy 100, no. 5 (1992): 992–1026.

29. Philip E. Tetlock, Expert Political Judgment: How Good Is It? How Can We Know? (Princeton: Princeton University Press, 2005).

30. Amos Tversky and Daniel Kahneman, "Judgment Under Uncertainty: Heuristics and Biases," Science 185, no. 4157 (1974): 1124–1131.

Chapter 9: Knowing Versus Accessing

31. Daniel T. Willingham, Why Don't Students Like School? (San Francisco: Jossey-Bass, 2009).

32. George A. Miller, "The Magical Number Seven, Plus or Minus Two," Psychological Review 63, no. 2 (1956): 81–97.

33. Nicholas Carr, The Shallows: What the Internet Is Doing to Our Brains (New York: W.W. Norton & Company, 2010).

34. Plato, Phaedrus, trans. Robin Waterfield (Oxford: Oxford University Press, 2002).

Chapter 10: Analysis Versus Judgment

35. Daniel Kahneman, Thinking, Fast and Slow (New York: Farrar, Straus and Giroux, 2011).

36. Gerd Gigerenzer, Risk Savvy: How to Make Good Decisions (New York: Viking, 2014).

37. Nassim Nicholas Taleb, The Black Swan: The Impact of the Highly Improbable (New York: Random House, 2007).

38. Philip E. Tetlock and Dan Gardner, Superforecasting: The Art and Science of Prediction (New York: Crown Publishing, 2015).

Chapter 11: The Amplified Mind

39. Andy Clark and David Chalmers, "The Extended Mind," Analysis 58, no. 1 (1998): 7–19.

40. Herbert A. Simon, Models of Bounded Rationality (Cambridge, MA: MIT Press, 1982).

41. Thomas W. Malone, Robert Laubacher, and Chrysanthos Dellarocas, "The Collective Intelligence Genome," MIT Sloan Management Review 51, no. 3 (2010): 21–31.

42. Douglas C. Engelbart, "Augmenting Human Intellect: A Conceptual Framework," Stanford Research Institute (1962).

Chapter 12: Designing Amplified Workflows

43. Herbert A. Simon, Administrative Behavior, 4th ed. (New York: Free Press, 1997).

44. Richard H. Thaler and Cass R. Sunstein, Nudge: Improving Decisions About Health, Wealth, and Happiness (New Haven: Yale University Press, 2008).

45. Gary Klein, Sources of Power: How People Make Decisions (Cambridge, MA: MIT Press, 1998).

46. Donald A. Norman, The Design of Everyday Things (New York: Basic Books, 2013).

Chapter 13: Designing Organizational Intelligence

47. Peter M. Senge, The Fifth Discipline: The Art and Practice of the Learning Organization (New York: Doubleday, 1990).

48. James G. March, "Exploration and Exploitation in Organizational Learning," Organization Science 2, no. 1 (1991): 71–87.

49. Thomas W. Malone, Superminds: The Surprising Power of People and Computers Thinking Together (New York: Little, Brown and Company, 2018).

50. Amy C. Edmondson, The Fearless Organization (Hoboken: Wiley, 2018).

Chapter 14: The Discipline of Amplified Thinking

51. Daniel Kahneman, Thinking, Fast and Slow (New York: Farrar, Straus and Giroux, 2011).

52. Anders Ericsson and Robert Pool, Peak: Secrets from the New Science of Expertise (Boston: Houghton Mifflin Harcourt, 2016).

53. Andy Clark, Supersizing the Mind: Embodiment, Action, and Cognitive Extension (Oxford: Oxford University Press, 2008).

54. Hannah Arendt, Responsibility and Judgment (New York: Schocken Books, 2003).

GLOSSARY

Terms are defined as they are used in this book.

Amplification

The use of AI systems to expand what human cognition can achieve without replacing the generative thinking that builds expertise and judgment. Amplification occurs when the human thinker generates the initial structure of a problem and uses the system to expand, challenge, and test that reasoning. Distinguished from substitution, in which the system generates the structure and the human evaluates it.

Amplified Mind Model

The framework introduced in Chapter 11 describing how human generation, knowledge compression, judgment, and AI-assisted analysis form an interdependent cycle. Generation builds internalized knowledge. Knowledge compression enables judgment. Judgment governs analysis. Analysis deepens generation. When any element is removed, the cycle degrades toward substitution.

Authority Transfer

The shift that occurs when a professional moves from originating reasoning to approving system-generated reasoning. Authority transfer is gradual and often invisible in the moment. It becomes visible when reasoning must be defended under pressure and the professional discovers the internal structure of the argument was never theirs.

Authoring versus Editing

The governing distinction of Chapter 6. Authoring generates the internal structure of an argument. Editing modifies a structure that already exists. AI workflows make it possible to edit endlessly without ever authoring, producing a professional who can present reasoning fluently but cannot reconstruct it independently.

Cognitive Offloading

Reliance on external systems to perform mental tasks that would otherwise require internal reasoning, memory, or problem solving. Earlier tools offloaded storage and retrieval. Artificial intelligence offloads idea generation itself, the initial construction of explanations and frameworks, which is where expertise develops.

Collective Intelligence

An institutional state in which multiple ways of framing problems exist simultaneously, different teams approach questions through different analytical lenses, and judgment remains distributed across people rather than concentrated in system outputs. The deliberate alternative to synchronized thinking.

Deliberate Practice

A specific form of effortful learning identified by psychologist Anders Ericsson that produces expertise. Requires three conditions: tasks at the edge of current capability, feedback that reveals failure, and an environment where failure is possible. AI systems tend to eliminate all three conditions simultaneously.

Desirable Difficulties

Conditions identified by cognitive psychologist Robert Bjork that make learning feel harder in the short term but strengthen retention and transfer over time. Spacing, retrieval practice, and interleaving are the primary examples. AI systems remove desirable difficulties by delivering answers before the productive struggle that builds understanding.

Distributed Generation

An organizational design principle introduced in Chapter 13. In institutions that preserve collective intelligence, the initial framing of important problems occurs across multiple independent perspectives before AI systems are consulted. Ensures the organization begins with multiple ways of seeing a problem rather than a single system-generated frame.

Fluency Distortion

The governing concept of Chapter 1. Occurs when the ease with which an explanation can be processed is mistaken for evidence that the explanation is correct. Clarity and correctness are not the same variable. AI systems produce high fluency regardless of the depth of reasoning behind the output, amplifying this bias beyond its historical range.

Generative Effort

The form of cognitive effort that builds internal capability rather than simply completing tasks. Occurs when the mind constructs understanding rather than applies it, organizing ideas, testing interpretations, discovering relationships between concepts. Distinguished from productive effort, which finishes work without necessarily building expertise.

Generative Thinking

Thinking that constructs ideas internally. Distinguished from reactive thinking, which responds to external inputs.

Generative thinking requires sustained, uninterrupted attention and is the cognitive state in which expertise forms.

Google Effect

A phenomenon identified by psychologist Betsy Sparrow and colleagues: people are less likely to remember information when they believe it remains digitally accessible. Instead of storing the information internally, the brain stores a pathway for retrieving it. A precursor to the broader pattern of cognitive offloading.

Illusion of Explanatory Depth

A cognitive phenomenon identified by psychologists Leonid Rozenblit and Frank Keil. People consistently overestimate how well they understand things they can recognize but cannot explain from first principles. AI-generated reasoning amplifies this illusion because familiarity with a generated argument feels like comprehension of it.

Informational Cascade

A dynamic identified by Cass Sunstein and colleagues in which individuals adopt beliefs because they observe others accepting them. AI systems create a new form of cascade that does not require social observation: when many professionals consult the same system, their reasoning converges analytically before any social interaction occurs.

Institutional Convergence

The organizational-scale version of authority transfer. When many professionals rely on the same AI systems, their analytical frameworks converge. The organization appears to maintain diverse perspectives while actually applying the same reasoning lens. Distinguished from individual deference by scale and by the difficulty of detection.

Judgment

The human capacity to decide what to do when analytical patterns are incomplete. Acts when evidence runs out, asks questions analysis cannot answer, weighs consequences under uncertainty, and accepts that some decisions must be made before information is fully available. AI systems excel at analysis; human reasoning remains essential for judgment.

Knowledge Compression

The process by which large bodies of information become organized into compact mental models through internalization and practice. Expertise depends on knowledge compression. Cannot be built through retrieval alone; requires the generative effort of constructing explanations and solutions independently.

Knowing versus Accessing

The governing distinction of Chapter 9. Knowing means information is internalized and immediately available as the foundation for reasoning. Accessing means the information exists somewhere retrievable but must be retrieved before thinking can begin. Under time pressure and uncertainty, knowing enables immediate reasoning; accessing requires a pause the situation may not permit.

Noise versus Signal

The governing distinction of Chapter 7. Protecting attention from noise, trivial distraction, requires discipline. Protecting attention from signal, useful AI output, requires judgment, because signal is difficult to resist precisely because it might contain something valuable.

Productive Effort

The effort required to complete tasks, writing a report, summarizing a document, producing an answer. AI systems reduce productive effort dramatically. Distinguished from generative effort, which builds capability rather than simply finishing work.

Reactive Thinking

Thinking that responds to external inputs rather than constructing ideas internally. Edits, selects, approves, or rejects. A necessary cognitive mode but insufficient on its own for building the reasoning structures that expertise depends on.

Retrieval versus Generation

The governing distinction of Chapter 2. Earlier tools changed where information was stored. AI changes where ideas come from. When ideas originate externally through generation, the internal processes that build expertise through the effort of constructing explanations occur less frequently.

Structured Dissent

An organizational design principle introduced in Chapter 13. After an AI-assisted analysis is produced, specific individuals or teams are assigned the role of questioning it, not refining it. Prevents machine-generated frameworks from becoming default truths within organizations.

Substitution

The use of AI systems to replace cognitive work that would otherwise build expertise and judgment. Appears efficient in the moment but gradually erodes the internal structures that support independent reasoning. Distinguished from amplification, which expands human cognition without replacing its generative source.

Synchronized Thinking

An organizational state in which the same analytical lens is applied across an institution, producing the appearance of diverse perspectives while the underlying reasoning has already converged. Efficient but dangerous under uncertainty, because when the shared framework is wrong, the error is distributed across the entire organization simultaneously.

Synthetic Fluency

The quality of AI-generated language that resembles well-formed reasoning regardless of whether the underlying logic is complete or correct. Persuasive because human judgment has always associated fluency with careful thought. AI produces fluency before reasoning has been tested.

Transactive Memory

A concept developed by psychologist Daniel Wegner describing how individuals in collaborative environments distribute knowledge across the group. AI systems increasingly occupy a similar role, with the critical difference that AI generates ideas rather than merely retrieving stored information.

INDEX

(This index was compiled utilizing AI, and verified by the Author)

Entries marked "Note" refer to the endnote on that page.

A

Accountability for judgment
  design principle, 141, 146–151, 157
  preserving human responsibility, 146–147, 150

Amplification
  defined, 122
  versus substitution, 119, 122–123, 125–126, 154–155
  workflow design for, 129–139
  at institutional scale, 148–149

Amplified Mind Model
  introduced, 126
  four-element cycle, 125
  scaled to institutions, 144

Analysis versus judgment
  governing distinction, 110
  where analysis ends, 111
  illusion of complete analysis, 112–113
  judgment governing analysis, 109–117

Arendt, Hannah–responsibility and judgment, 167 Note 54

Attention
  as precondition of authorship, 84–85, 87
  discipline of delay, 85–86
  fragmentation of, 55
  noise versus signal, 82–83

Authoring versus editing
governing distinction, 72
procedural shift, 74
responsibility problem, 75
illusion of explanatory depth and, 73

Authority transfer
defined, 44
bridge-and-inspector metaphor, 45
diagnostic question, 47
at institutional scale, 89–90, 92, 94–96

B

Bikhchandani, Sushil–informational cascades, 164 Note 28

Bjork, Elizabeth L.–desirable difficulties, 163 Note 18

Bjork, Robert–desirable difficulties, 65, 163 Note 18

Blind spots, institutional, 93, 95

Bridge-and-inspector metaphor, 45

C

Carr, Nicholas–The Shallows, 13, 165 Note 33

Chalmers, David. See Extended Mind (Clark and Chalmers)

Clark, Andy
Extended Mind (with Chalmers), 166 Note 39
Supersizing the Mind, 167 Note 53

Cognitive offloading
defined, 34
storage versus generation distinction, 34, 36
Google Effect, 34–35
schemas and germane load, 36–37
compounds with fluency distortion, 40
accelerated by attention fragmentation, 55

Collective intelligence
defined, 143

versus synchronized thinking, 141–149, 151
three design principles, 144–146, 148–151

Csikszentmihalyi, Mihaly–flow states, 164 Note 27

D

Darley, John M. and Latané, Bibb–diffusion of responsibility, 163 Note 20

Deep work (Newport), 13, 55, 82, 163 Note 15, 164 Note 25

Deliberate practice (Ericsson), 38, 64–65, 161 Note 7, 163 Note 17, 167 Note 52

Dellarocas, Chrysanthos–collective intelligence, 166 Note 41

Desirable difficulties (Bjork), 65, 163 Note 18

Dietvorst, Berkeley J.–algorithm aversion, 162 Note 9

Discipline of delay, 85–86

Distributed generation, 144–145, 149, 157

E

Edmondson, Amy C.–the fearless organization, 167 Note 50

Engelbart, Douglas–augmenting human intellect, 166 Note 42

Ericsson, Anders–deliberate practice, 38, 64–65, 161 Note 7, 163 Note 17, 167 Note 52

Expertise atrophy, 66–67

Extended Mind (Clark and Chalmers), 166 Note 39

F

Fast intuitive acceptance, 28

Feynman Technique, 78, 164 Note 23

Financial crisis (2008)–analytical convergence parallel, 93

Fluency distortion
defined, 26
processing fluency research, 26–28
three interacting forces, 28

compounds with cognitive offloading, 40
Friction–cognitive role of, 27–32, 65–66, 122–126, 155–156

G

Gardner, Dan–superforecasting, 165 Note 38
Generation before retrieval, 123, 132–133, 156. See also Distributed generation
Generative effort
defined, 62–63
versus productive effort, 61–70
beyond Chapter 5, 30, 38, 74, 77
sustained in amplified workflows, 153–156, 158, 160
Generative thinking
defined, 52
versus reactive thinking, 52–58, 84, 86
Gigerenzer, Gerd–risk savvy, 165 Note 36
Gleick, James–Genius (Feynman), 164 Note 23
Gonzalez, Victor M.–fragmented work, 162 Note 14
Google Effect (Sparrow), 34–35, 161 Note 4
Governing distinctions, summarized, 154. See also Glossary, 169–175

H

Harris, Justin–fragmented work, 162 Note 14
Hirshleifer, David–informational cascades, 164 Note 28

I

Illusion of complete analysis, 112–113
Illusion of explanatory depth (Rozenblit and Keil), 73, 163 Note 21
Informational cascades (Sunstein), 91–92, 164 Note 28
Institutional convergence
defined, 89

2008 financial crisis parallel, 93
diversity of voice, uniformity of thought, 94
recurrence, 92, 96, 115
design alternatives, 144–146

Intentional versus accidental design, 129, 131–132, 136, 138–139

J

Judgment
nature of, 112
four questions of, 112, 116–117
decision-making without precedent, 113–114
checkpoints in workflow, 134–135, 138, 156
accountability for, 146–147, 150, 157

K

Kahneman, Daniel, 13, 161 Note 2, 162 Note 11, 165 Note 30, 165 Note 35, 167 Note 51

Keil, Frank. See Rozenblit, Leonid and Keil, Frank

Klein, Gary–naturalistic decision-making, 163 Note 19, 166 Note 45

Knowing versus accessing
governing distinction, 100
fragmented knowledge under pressure, 99–107
illusion of external intelligence, 104

Knowledge compression
defined, 102
grandmaster and firefighter examples, 102
built through generation, 103, 105
in Chapter 9, 104, 106–107
in the Amplified Mind Model, 112, 125, 154

L

Latané, Bibb. See Darley, John M. and Latané, Bibb

Laubacher, Robert–collective intelligence, 166 Note 41

Leroy, Sophie–attention residue, 162 Note 13
Lieberman, Daniel–The Story of the Human Body, 9 (Foreword)
Liu, Jenny–Google Effect, 161 Note 4
Logg, Jennifer M.–algorithm appreciation, 162 Note 10

M

Machine-generated fluency, 28
Malone, Thomas W.–collective intelligence, 166 Note 41, 167 Note 49
March, James G.–exploration and exploitation, 167 Note 48
Mark, Gloria–attention research, 162 Note 14, 164 Note 26
Massey, Cade–algorithm aversion, 162 Note 9
Miller, George–the magical number seven, 165 Note 32
Minson, Julia A.–algorithm appreciation, 162 Note 10
Moore, Don A.–algorithm appreciation, 162 Note 10

N

Naturalistic decision-making (Klein), 163 Note 19, 166 Note 45
Newport, Cal–deep work, 13, 55, 82, 163 Note 15, 164 Note 25
Noise versus signal
  governing distinction, 82, 85
  why signal is harder to resist, 83, 87
Norman, Donald–the design of everyday things, 166 Note 46

O

Originating versus approving
  governing distinction, 43, 46
  diagnostic question, 47
  approval as the default role, 48–49

P

Parasuraman, Raja and Riley, Victor–humans and automation, 161

Note 3, 162 Note 8

Plato
  on writing, 9 (Foreword)
  Phaedrus, 165 Note 34

Pool, Robert–Peak, 161 Note 7, 163 Note 17, 167 Note 52

Processing fluency
  research on, 26
  AI amplification of, 26–28

Productive versus generative effort
  governing distinction, 62–63, 68

R

Reactive thinking
  defined, 52
  versus generative thinking, 52–58
  Simon's attention poverty and, 52, 54, 57–58

Retrieval versus generation
  governing distinction, 34, 36

Riley, Victor. See Parasuraman, Raja and Riley, Victor

Rozenblit, Leonid and Keil, Frank–explanatory depth, 73, 163 Note 21

S

Schemas (Sweller)
  formation of, 36–38
  germane cognitive load, 37

Schwarz, Norbert–metacognitive experiences, 161 Note 1

Senge, Peter M.–the fifth discipline, 166 Note 47

Simmons, Joseph P.–algorithm aversion, 162 Note 9

Simon, Herbert–wealth of information creates poverty of attention, 52, 54, 57–58, 162 Note 12, 164 Note 24, 166 Note 40, 166 Note 43

Sparrow, Betsy–Google Effect, 34–35, 161 Note 4

Structured dissent, 145–146, 148, 150, 157

Substitution versus amplification
governing distinction, 122
why substitution is easy, 122–123
why amplification requires discipline, 123

Sunstein, Cass–informational cascades, 91, 164 Note 28, 166 Note 44

Sweller, John–cognitive load theory, 36–37, 161 Note 6, 163 Note 16

Synchronized thinking
defined, 143
versus collective intelligence, 141, 143–149, 151

Synthetic fluency
defined, 15, 17–18
See also Fluency distortion

T

Taleb, Nassim Nicholas–black swan events, 165 Note 37

Tetlock, Philip–expert judgment, 165 Note 29, 165 Note 38

Thaler, Richard H.–nudge, 166 Note 44

Transactive memory (Wegner), 35, 161 Note 5

Tversky, Amos–heuristics and biases, 165 Note 30

W

Wegner, Daniel–transactive memory, 35, 161 Note 4, 161 Note 5, 163 Note 22

Welch, Ivo–informational cascades, 164 Note 28

Wildfire commander scene, 109–110

Willingham, Daniel–why knowledge matters, 165 Note 31

Workflow design
individual, 129–139
organizational, 144–148
accidental versus intentional, 129, 131, 139

ABOUT THE AUTHOR

Paul F. Alden served more than 38 years in the fire service and retired as a Fire Captain with the Plymouth Fire Department. He became a firefighter and emergency medical technician in 1985, and for nearly four decades he honed his skills, recertified every year, and worked a frontline 911 system, the kind of environment where decisions get made with incomplete information and real consequences.

It was in that work that he first saw the pattern this book is about. Emergency medicine had moved toward decision trees and standardized protocols, and for good reasons. They save lives and they reduce error. But over time he watched something quieter happen. The protocols that were meant to support judgment began, in some hands, to replace it. The skills required to actually diagnose a patient, to reason from symptoms back to cause rather than match an input to a branch, were no longer being exercised, and they began to atrophy. In recent years the field recognized this and began teaching diagnostic reasoning again, deliberately, as a discipline that has to be maintained rather than assumed. He came to believe the same thing is now happening with artificial intelligence, only far faster, and that recognizing it in time is the reason he wrote this book.

Captain Alden approached the book the way he approached his career. In the fire service you do not bring a problem to someone's attention without at least a possible solution. So the book does both. It diagnoses the problem first, and then it builds a practical response.

He lives between the Boston area of Massachusetts and the Hilton Head area of South Carolina, depending on the season, with his wife, Leslie, and their two children, AJ and Delaney.

What he hopes readers take away is simple. The machine can generate the answer. Whether the reader remains the person who can think it through is a skill, and like any skill, it is kept only by using it.

ABOOKS

ALIVE Book Publishing and ALIVE Publishing Group
are imprints of Advanced Publishing LLC,
3200 A Danville Blvd., Suite 204, Alamo, California 94507

Telephone: 925.837.7303
alivebookpublishing.com